Lecture Notes in Networks and Systems

Volume 1700

Series Editor

Janusz Kacprzyk ⓘ, Systems Research Institute, Polish Academy of Sciences, Warsaw, Poland

Advisory Editors

Fernando Gomide, Department of Computer Engineering and Automation—DCA, School of Electrical and Computer Engineering—FEEC, University of Campinas—UNICAMP, São Paulo, Brazil

Okyay Kaynak, Department of Electrical and Electronics Engineering, Bogazici University, Istanbul, Türkiye

Derong Liu, Department of Electrical and Computer Engineering, University of Illinois at Chicago, Chicago, USA
   Institute of Automation, Chinese Academy of Sciences, Beijing, China

Witold Pedrycz, Department of Electrical and Computer Engineering, University of Alberta, Edmonton, Alberta, Canada
   Systems Research Institute, Polish Academy of Sciences, Warsaw, Poland

Marios M. Polycarpou, Department of Electrical and Computer Engineering, KIOS Research Center for Intelligent Systems and Networks, University of Cyprus, Nicosia, Cyprus

Imre J. Rudas, Óbuda University, Budapest, Hungary

Jun Wang, Department of Computer Science, City University of Hong Kong, Kowloon, Hong Kong

The series "Lecture Notes in Networks and Systems" publishes the latest developments in Networks and Systems—quickly, informally and with high quality. Original research reported in proceedings and post-proceedings represents the core of LNNS.

Volumes published in LNNS embrace all aspects and subfields of, as well as new challenges in, Networks and Systems.

The series contains proceedings and edited volumes in systems and networks, spanning the areas of Cyber-Physical Systems, Autonomous Systems, Sensor Networks, Control Systems, Energy Systems, Automotive Systems, Biological Systems, Vehicular Networking and Connected Vehicles, Aerospace Systems, Automation, Manufacturing, Smart Grids, Nonlinear Systems, Power Systems, Robotics, Social Systems, Economic Systems and other. Of particular value to both the contributors and the readership are the short publication timeframe and the world-wide distribution and exposure which enable both a wide and rapid dissemination of research output.

The series covers the theory, applications, and perspectives on the state of the art and future developments relevant to systems and networks, decision making, control, complex processes and related areas, as embedded in the fields of interdisciplinary and applied sciences, engineering, computer science, physics, economics, social, and life sciences, as well as the paradigms and methodologies behind them.

Indexed by SCOPUS, EI Compendex, INSPEC, WTI Frankfurt eG, zbMATH, SCImago.

All books published in the series are submitted for consideration in Web of Science.

For proposals from Asia please contact Aninda Bose (aninda.bose@springer.com).

Dirk Vanhoudt
Editor

Proceedings of the 19th International Symposium on District Heating and Cooling

The International Research Conference on Heating and Cooling Networks—Under the Supervision of IEA DHC

 Springer

Editor
Dirk Vanhoudt
VITO/EnergyVille
Mol/Genk, Belgium

ISSN 2367-3370 ISSN 2367-3389 (electronic)
Lecture Notes in Networks and Systems
ISBN 978-3-032-09843-6 ISBN 978-3-032-09844-3 (eBook)
https://doi.org/10.1007/978-3-032-09844-3

This work was supported by Dirk Vanhoudt.

© The Editor(s) (if applicable) and The Author(s) 2026. This book is an open access publication.

Open Access This book is licensed under the terms of the Creative Commons Attribution 4.0 International License (http://creativecommons.org/licenses/by/4.0/), which permits use, sharing, adaptation, distribution and reproduction in any medium or format, as long as you give appropriate credit to the original author(s) and the source, provide a link to the Creative Commons license and indicate if changes were made.

The images or other third party material in this book are included in the book's Creative Commons license, unless indicated otherwise in a credit line to the material. If material is not included in the book's Creative Commons license and your intended use is not permitted by statutory regulation or exceeds the permitted use, you will need to obtain permission directly from the copyright holder.

The use of general descriptive names, registered names, trademarks, service marks, etc. in this publication does not imply, even in the absence of a specific statement, that such names are exempt from the relevant protective laws and regulations and therefore free for general use.

The publisher, the authors and the editors are safe to assume that the advice and information in this book are believed to be true and accurate at the date of publication. Neither the publisher nor the authors or the editors give a warranty, expressed or implied, with respect to the material contained herein or for any errors or omissions that may have been made. The publisher remains neutral with regard to jurisdictional claims in published maps and institutional affiliations.

This Springer imprint is published by the registered company Springer Nature Switzerland AG
The registered company address is: Gewerbestrasse 11, 6330 Cham, Switzerland

If disposing of this product, please recycle the paper.

Proceedings of
the 19th International Symposium on
District Heating and Cooling

Editor: Dirk Vanhoudt – VITO/EnergyVille

Preface

The *19th International Symposium on District Heating and Cooling* was held from 7 to 10 September 2025 in Genk, Belgium. It was organised by VITO/EnergyVille under the umbrella of IEA DHC, the Technology Collaboration Programme on District Heating and Cooling of the International Energy Agency (IEA).

Since its first edition in Lund, Sweden, in 1987, the Symposium has established itself as the premier scientific forum for district heating and cooling research. While originally rooted in the Nordic and Central European research landscape, the Symposium has evolved into a truly global platform since 2016, welcoming participants and contributions from across all continents. The IEA DHC Symposium continues to foster international collaboration and provide a space for the exchange of knowledge and innovative solutions to the pressing technological, environmental, and societal challenges facing the sector.

The 2025 edition featured four keynote speakers from different continents, each offering a regional perspective on DHC developments. Together, their contributions provided a global overview of current trends, challenges, and innovations in the sector across Europe, Asia, and North America. The programme offered a broad spectrum of peer-reviewed research covering the following topics: the digitalisation of DHC networks, integration of diverse thermal sources, system sustainability, user engagement, geothermal heating and cooling, cost-optimal design, and modelling and data handling. Special attention was given to promising contributions by early-career researchers, with the DHC+ Student Awards highlighting outstanding bachelor's, master's, and Ph.D. work.

The papers presented in this volume represent a limited selection from the full conference programme contributions. Each paper was peer-reviewed by members of the Scientific Committee and evaluated for scientific quality, relevance, and clarity. Together, they offer insight into the current state of the field and the directions in which global DHC research and practice are evolving.

As editor, I thank the International Energy Agency (IEA) and the IEA DHC TCP for their continued support and Springer Nature for enabling open access to this selection of abstracts. Special thanks also go to all authors and Scientific Committee members for their valuable contributions. This publication aims to serve as a valuable reference and source of inspiration for all those committed to advancing sustainable heating and cooling systems worldwide.

VITO/EnergyVille
September 2025

Dirk Vanhoudt

Organization

Conference Chairs

Ann Wouters, Research Program Manager Digital Solutions for Energy Systems at VITO/EnergyVille, Belgium
Dirk Vanhoudt, Research Roadmap Responsible Digitalization of District Heating and Cooling Systems at VITO/EnergyVille, Belgium

Organizing Committee

Ann Wouters, Research Program Manager Digital Solutions for Energy Systems at VITO/EnergyVille, Belgium
Dirk Vanhoudt, Research Roadmap Responsible Digitalization of District Heating and Cooling Systems at VITO/EnergyVille, Belgium
Erik De Schutter, Business Developer Thermal Energy Systems at VITO/EnergyVille, Belgium
Adinda Vandereyken, Event Organiser at EnergyVille, Belgium
Nathalie Belmans, Event Organiser at KU Leuven/EnergyVille, Belgium

Scientific Committee

Michael Ahern, Ever-green Energy, USA
Prof. Markus Blesl, University of Stuttgart, Germany
Prof. Maarten Blommaert, KU Leuven/EnergyVille, Belgium
Stef Boesten, Rijksdienst voor Ondernemend Nederland, the Netherlands
Raymond Boulter, National Resources Canada, Canada
Casey Collins, Duke University, USA

Dr. Alice Dénarié, Politecnico di Milano, Italy
Prof. Elisa Guelpa, Politecnico di Torino, Italy
Dr. Aleksandr Hlebnikov, Tallinn University of Technology, Estonia
Prof. Stefan Holler, HAWK University of Applied Sciences and Arts, Germany
Prof. Anton Ianakiev, Nottingham Trent University, UK
Peter Kahoe, Sustainable Energy Authority of Ireland, Ireland
Dr. Hanne Paulina Kauko, Sintef, Norway
Håkan Knutsson, SweHeat and Cooling, Sweden
Prof. Risto Kosonen, Aalto University, Finland
Dr. Lukas Kranzl, Technical University Vienna, Austria
Prof. Bruno Lacarrière, IMT Atlantique, France
Prof. Henrik Lund, Aalborg University, Denmark
Prof. Natasa Nord, NTNU, Norway
Dr. Miika Rama, VTT Technical Research Centre of Finland, Finland
Dr. Étienne Saloux, National Resources Canada, Canada
Dr. Ralf-Roman Schmidt, Austrian Institute for Technology, Austria
Prof. Sylvain Serra, University of Pau, France
Prof. Eva Thorin, Mälardalen University, Sweden
Prof. Michele Tunzi, Danish Technical University, Denmark
Dirk Vanhoudt, VITO/EnergyVille, Belgium
Prof. Anna Volkova, Tallinn University of Technology, Estonia
Prof. Haichao Wang, Dalian University of Technology, China
Dr. Robin Wiltshire, BRE, UK
Prof. Jianjun Xia, Tsinghua University, China

Contents

Modeling with the 223P Ontology and Adaptation for Optimized DHC Operation 1
Steffen Wallner, Mathias Ziebarth, and Christian Wolff

Optimizing a Multi-vector Energy System with Geothermal-Powered District Heating 11
Natalia Kozlowska, Arthur Lefebvre, Julien Jacquemin, and Pierre Dewallef

Design Optimization of District Heating Networks with Thermal-Hydraulic Validation 21
Yacine Gaoua, Charlie Pretot, and Nicolas Vasset

Large-Scale Solar Thermal Systems in District Heating Networks: A Review of German Projects Regarding Dimensioning, Temperatures and Stagnation Times 31
Bert Schiebler, Maik Kirchner, Julian Jensen, and Federico Giovannetti

Enabling Data Economy for DHC Applications Through Dataspaces 41
Edmund Widl, Christian Wolff, Mathieu Vallée, Kristina Lygnerud, Dietrich Schmidt, Zheng Grace Ma, Xiaochen Yang, Axel Oliva, and Michele Tunzi

Mitigating the Grid Impact of Electrifying Heat: A Case Study Using Solar and Thermal Storage to Achieve a Net-Zero Emission Campus 51
Eric Ho, Raymond Boulter, and Jeff Thornton

Natural Circulation and Other Measures to Ensure Heating Supply to Buildings Connected to District Heating in the Event of Electrical Grid Blackout 63
Merilin Nurme, Dabrel Prits, Karl-Villem Võsa, Andrei Dedov, and Anna Volkova

Feasibility Assessment Tool for District Heating and Cooling (FAST DHC): A Simple Decision Support Tool for the Techno-Economic Evaluation of DHC Networks 75
Henrique Lagoeiro, Nicolas Marx, Alessandro Maccarini, Oddgeir Gudmundsson, Catarina Marques, Ralf-Roman Schmidt, and Graeme Maidment

Testing Dynamic Sector-Coupled Operation in a District Energy Laboratory 89
Oliver Gehrke, Lucas Venge Ejlsborg, and Kai Heussen

TFSB as Bedding Material in District Heating Pipe Construction–Scientifically Proven Long-Term Experience 99
Bernd Wagner, Stefan Hay, Thomas Neidhart, Florian Spirkl, Michael Ried, Louis Zrenner, Ingo Weidlich, Eugen Gabriel, and Timo Banning

Experimental Investigation on the Thermal Conductivity of Alternative Backfill Materials for District Heating Networks 109
Stefan Dollhopf and Ingo Weidlich

Achieving Efficient District Heating Targets in a Croatian Network: Heat Source Mapping and Techno-Economic Scenarios Analysis 119
Daniele Anania, Josip Miškić, Tomislav Pukšec, and Marco Cozzini

Exploiting Synergies of Data-Driven and Model-Based Approaches for Leakage Localization in District Heating Networks: Application of Improved Approaches 133
Dennis Pierl, Julia Koltermann, Kai Vahldiek, Bernd Rüger, Kai Michels, Andreas Nürnberger, and Frank Klawonn

Data-Driven Fault Detection and Diagnosis in District Heating Substations and the Impact of Return Temperature Reduction 143
Vera Alieva, Tiedo Behrends, Vera Boß, Peter Stange, and Clemens Felsmann

Data Pre-processing Methods Enhancing Heat Cost Allocator Measurement Usability 153
Qinjiang Yang, Fabio Saba, Marina Orio, Marco Santiano, Emanuele Audrito, Robbe Salenbien, and Michele Tunzi

Foam Density Distribution Analysis in Pre-insulated Pipes Using Non-destructive X-Ray Microscopy 163
Pakdad Langroudi and Ingo Weidlich

A 5th Generation District Heating and Cooling Network (5GDHC) Driven by Shallow Geothermal, Economic Analysis and Geohydrological Modelling for Comparison with an Individual System 173
Hans Hoes and Jente Pauwels

Improved District Heating Return Temperatures by Cascading Concepts 183
Jan Eric Thorsen, Oddgeir Gudmundsson, Michele Tunzi, and Marek Brand

Status of the VITO Deep Geothermal Project in Mol—Donk (Northern Belgium) 193
Ben Laenen and Matsen Broothaers

Proposal for a Method to Simultaneously Maximize the Economic and Environmental Values of a District Heating and Cooling System for an Electricity Market 203
Kohei Tomita, Yutaka Iino, and Yasuhiro Hayashi

Optimizing the Next Generation of District Heating and Cooling Systems While Ensuring Reliable Domestic Hot Water Supply 213
Mohammad-Reza Kolahi and Martin K. Patel

Closing a Sim-to-Sim Gap for Automatic Fault Detection in DHC Systems Using Hybrid Modelling 223
Pieter Jan Houben, Stef Jacobs, Renzo Massobrio, Ivan Verhaert, and Peter Hellinckx

Modeling with the 223P Ontology and Adaptation for Optimized DHC Operation

Steffen Wallner, Mathias Ziebarth, and Christian Wolff

Abstract District heating and cooling (DHC) offers great potential for efficient, cost-effective and flexible large-scale use of low-carbon energy in the building sector. However reaching Net Zero Emissions by 2050 requires significant efforts to rapidly improve the energy efficiency of existing networks, switch them to renewable heat sources, integrate secondary heat sources and to develop new high-efficiency infrastructure. A critical factor for reaching these goals is the digitalization of these systems to access new tools like the identification of problematic consumer substations, predictive maintenance, optimized heat production, evaluation of potential extensions and efficient sector coupling. This paper proposes an extension of the emerging ASHRAE 223P standard for machine-readable semantic frameworks representing building automation and control data, and other building system information to DHC networks. it is validated through two use cases in two distinct research projects. The adapted 223P ontology could be a starting point for the development of a standard data model of DHC networks.

Keywords District heating networks · Digitalization · Ontology · Standardization · District heating and cooling

---

S. Wallner (✉) · M. Ziebarth
Fraunhofer Institute of Optronics, System Technologies and Image Exploitation (IOSB), Karlsruhe, Germany
e-mail: steffen.wallner@iosb.fraunhofer.de

M. Ziebarth
e-mail: mathias.ziebarth@iosb.fraunhofer.de

C. Wolff
Fraunhofer Institute for Solar Energy Systems (ISE), Freiburg, Germany
e-mail: christian.wolff@ise.fraunhofer.de

1 Introduction

The District heating and cooling (DHC) domain is critical for advancing a sustainable, emission-free heating sector. The transformation and expansion of district heating networks (DHN) necessitate climate-neutral generation, efficient operational strategies, and decentralized energy feed-in, which introduce complexities in management. This demands rigorous monitoring, real-time simulations, leak detection, and optimized heat production scheduling, all of which rely on substantial and well-structured input data. The integration of remotely readable measuring devices will enable network operators to capture data with greater frequency and near real-time accuracy. A pivotal challenge remains in the effective utilization of this data in conjunction with other available datasets across diverse applications. Due to the absence of a uniform standard, network operators either create their own (incompatible, short-lived) data models or depend on service providers, potentially resulting in vendor lock-in. To effectively utilize software applications, measurement data must be associated with metadata. This is frequently accomplished manually by translating data into application-specific formats and linking disparate datasets. Such data translations are often time-consuming and costly. According to Prakash et al. [1] a similar problem arises in the building sector, where between 20 and 29% of total implementation time is spent identifying and connecting the inputs and outputs of model predictive control controllers to the correct data sources and control points. They investigate ontologies (BRICK [2], 223P [3] and SAREF [4]) that could lead to plug-and-play development and deployment of new MPC controllers for efficient building operation. The same would be preferable for DHC.

Therefore, we employ an ontology-based modeling approach that stores all relevant information within what is referred to as an integrated data model. The principal advantage is the provision of a unified interface for the implementation of advanced applications as described in the use cases. Given that the information originates from various domains, it is modeled using different ontologies. The core of an integrated data model thus comprises:

1. The completeness and consistency of all stored information.
2. The connections between information from different domains.

Due to this approach and the described problems, this paper analyses the requirements of the DHC sector to an ontology and the suitability of existing ontologies as a core ontology for an extension. Based on this the developed extension of the 223P ontology for the district heating sector is introduced. This extension aims to incorporate additional properties and relationships relevant to DHC, thereby enhancing its applicability and facilitating data exchange among various stakeholders. It is validated in case studies from two research projects.

2 Related Work

To gain an overview of existing ontologies and data models, a variety of ontologies from energy system modeling were collected and assessed for their suitability. The focus is on system modeling ontologies that already encompass certain aspects necessary for inclusion in a DHC data model and could be used as core ontology. A key feature of a DHC network is its topology, where the piping system must be connected in a bidirectional manner to accommodate shifting mass flows within the network. Additionally, heat transfer stations, measurement devices, and production units also need to be incorporated. These aspects should at least be included in the concepts of a potential core ontology. The ability to model buildings or power systems is beneficial for sector coupling. Furthermore, important aspects of effective ontologies that must be considered include availability, maintenance, and alignment with other ontologies. For a detailed literature review we want to refer the reader to a recent preprint by Zeng Peng et al. [5] which identifies weaknesses of existing ontologies for the application in the DHC sector.

Table 1 presents promising ontologies, reviewed alongside a brief evaluation of their suitability for use or extension. Only actively maintained or standardized ontologies have been considered. Therefore, the "EE-district" [6], "Seas Ontology" [7], "Flow System Ontology" (FSO [8]), "Tubes system ontology" (TSO [9]) and "Sense" [10] are not relevant for extensions, although some features or ideas from these ontologies could be adapted. The "Open Energy Ontology" (OEO [11]) is designed for top-level system simulation and analysis, providing basic concepts for producers and heat transfer stations. However, it lacks detailed attributes and does not include a mapping for the network topology, specifically the piping system. In contrast, "Brick Schema" [2] focuses on the building sector and its assets. While it defines some core components, it does not address the piping system, nor does it provide for bidirectional connections between components. The "SAREF ontology suite" SAREF [4] offers a core ontology along with several extensions for various domains related to

Table 1 Comparison of ontologies with respect to necessary concepts in the DHC domain, availability, standardization and last updates.

Ontology, Aspect	FSO	223P	SAREF	TSO	EE-district	Seas	OEO	BRICK	Sense
Producers		✓	✓		✓	(✓)	✓	✓	✓
Consumers		✓	✓		✓	(✓)	✓	✓	
Sensors		✓	✓	✓	✓	✓	✓	✓	✓
Pipes	✓	✓	✓	✓	✓	(✓)			
Connections	✓	✓	✓	✓					
Availability	✓	✓	✓	✓		✓	✓	✓	✓
Standarized		(✓)	✓				(✓)	(✓)	
Up-to-date	2021	2025	2025	2022	2018	2017	2025	2025	2023

smart applications, including energy, water, and buildings. However, an extension specifically for district heating is lacking. Although the existing extensions cover many of the requirements for the DHC domain in concept form, further definition is needed. Overall, it appears suitable and warrants deeper exploration. Similarly, 223P [3] encompasses all requirements of the DHC domain within its concepts, but it lacks specific classes and concepts tailored to district heating. Nevertheless, alignments with building topology ontologies such as "RealEstateCore" and "Brick Schema" are feasible due to their similar core concepts.

Additionally, there are several data models and terminologies that can offer relevant concepts for a potential extension. These include the data models "CityGML Network Utility ADE" [12], "IFC 4.3" [13], as well as the terminologies "AGFW Dictionary" [14] and the "buildingSMART Dictionary" [15].

3 Selection of Base Ontology

Out of the two possible base ontologies discussed in Sect. 2, SAREF [4] and 223P [3], both are suitable candidates for extending the DHC domain. This paper focuses particularly on the bidirectional connections between pipes, emphasizing the importance of mapping a DHN topology to a data model. The following sections will highlight the differences between the ontologies to find a suitable core ontology for the DHC domain.

Both ontologies feature a class that facilitates connections to other entities. In SAREF, this is the "System" class from the "SAREF4SYST" extension, while in 223P, it is the "Connectable" class. Entities that are subclasses of these classes can be connected using similar properties, such as "connectedTo" or "connects." The core principles of both ontologies are thus quite similar, indicating that alignment between these concepts should be feasible. However, due to minor differences in their topological and connection concepts, further comparison between the two ontologies is necessary.

From a technical point of view 223P offers more restrictive validation rules than SAREF using closed world assumptions based on SHACL. This helps avoiding possible problems in the data and makes the ontology more suitable for practical use. Both ontologies provide concepts from the building domain, but at the time of writing 223P offers a significantly larger collection, especially taking into account the aligned ontologies REC[1] and BRICK [2].

In conclusion, 223P appears to be the more suitable core ontology for extending the DHC domain due to its stronger connections to the building domain and the advantages offered by its closed world assumption, which facilitates easier utilization of the ontology.

---

[1] https://github.com/RealEstateCore/realestatecore.github.io.

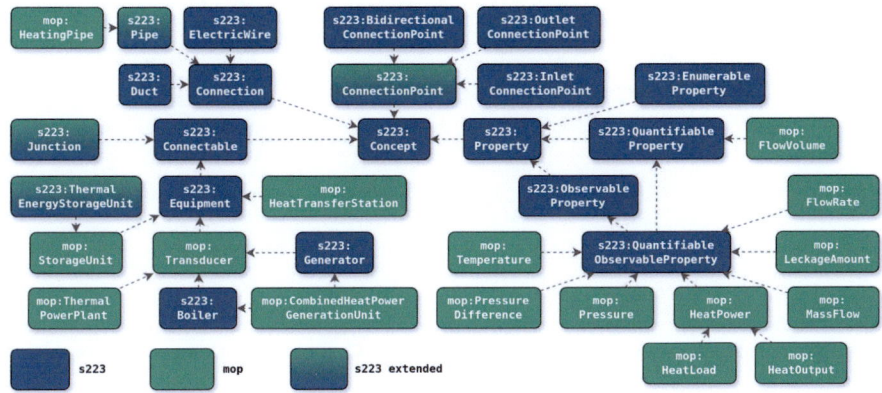

Fig. 1 Enhancing the 223P ontology with new and extending existing classes

4 Method

While 223P addresses the main requirements, it lacks specificity for the DHC domain and certain essential concepts that require extension. Especially we want our integrated data model to support thermo-hydraulic simulation models, geographical references, optimized operation and heat generation scheduling as well as references to measurement data. To fill these gaps, we have developed the MOP (Model for Operational Planning) ontology extensions for the 223P. The draft ontology is accessible online,[2] and its visualization employs a fork of the rdf-toolkit project.[3]

The deficiencies in 223P for DHC are missing concepts, classes and properties and insufficient specificity. For instance, it lacks a concept for determining the origin of a pipe's local coordinate system, crucial for simulation models. We introduced `mop:origin` to address this, pointing to a `s223:ConnectionPoint` as the origin.

The current 223P ontology although lacks specific classes for essential system components in the DHC domain. Existing classes are overly generic, compromising information quality. To address this, we introduced three new abstract classes that define crucial properties for readiness, performance, ramp-up times, emission, etc. and define connectivity constraints. From those we derived the classes shown in Fig. 1 that we integrated into the existing 223P taxonomy.

- `mop:HeatingPipe` extends `s223:Pipe` by adding thermal insulation details. The extended class now includes parameters for diameter and roughness.
- `mop:StorageUnit` describes maximum charging energy, charging and discharging functions and is an abstraction of the existing thermal storage.

---

[2] https://open.pages.fraunhofer.de/iosb/mrd/terms/#/mop/v1/.
[3] https://github.com/KrishnanN27/rdf-toolkit.

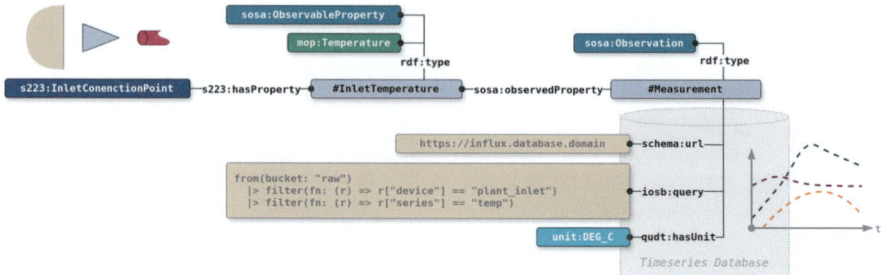

Fig. 2 On the left is the connection point of, for example, a system and a pipe. A temperature is measured at this point and there is an observation for this. The observation contains the location of the database, the database query and the unit

- `mop:ThermalPowerPlant` allows aggregation of generators, storage tanks, and pumps without detailed modeling of power plant hydraulics.

In general, the flexibility of 223P allows for the creation of technically implausible models within the DHC domain. One option would be to introduce further SHACL constraints. Despite the resulting increase in complexity, the MOP extension should remain understandable for users. For this reason, we have added modeling guidelines to the documentation. The drawback is that these guidelines are not automatically evaluable with SHACL.

4.1 Integration of Time Series Data

The implementation of algorithms in use cases necessitates not only relevant information but also its integration with time series data, such as measurements or synthetic data from forecast models. Additionally, dynamic price series for electricity and fuels or artificial price signals are required for certain use case.

The 223P framework includes a concept where a property points to an external reference, like the `ref:TimeseriesReference` in BRICK, which refers to a database table. However, this concept is not detailed, leaving specificity to the user. Alternatively, the SSN/SOSA[4] ontology offers a comprehensive approach for modeling observations and actuators. It consistently models relationship when measurement campaigns vary or when parameters, sensors or database systems change over time. Integrating time series data requires that the measurement unit must be part of the observation. Figure 2 illustrates an example of the connection between a temperature property and the stored time series data in an Influx database.

---

[4] https://www.w3.org/TR/vocab-ssn/.

Discussion is needed on whether SSN/SOSA concepts could be implemented in 223P. Aligning the ontologies is more sensible, preventing redundancies and complementing concepts. For example defining an s223:ObservableProperty as an sosa:ObservableProperty is a good starting point.

5 Use Cases

The approaches presented are used in two research projects to create and use integrated data models. Figure 3 gives an impression of the level of detail of the model and also shows some modeling guidelines that we have used.

First project SimKI-Mop focuses on optimizing district heating networks by enhancing the operation of decentralized power plants, considering transient network behavior. A key aspect is generating synthetic data sets for training data-driven models, which offer time advantages in optimization.

The integrated data model incorporates historical time series data stored in CSV files and databases mapped via SSN/SOSA to relevant properties. Additionally, the main components of the DHNs are detailed using the MOP ontology classes, enabling automatic creation of optimization models for the various decentralized power plant locations.

The second project 3SmartFW aims to detect leaks in the district heating network by analyzing pressure measurements online. Results of a leak detection and localization (LDL) system are visualized in real-time. To disconnect an affected subnetwork, the valves to be closed are automatically determined based on the network topology.

The integrated data model covers the pipe network, transfer stations, valves, their coordinates, and relationships. Online measurement data is stored in time series databases and sent to the web app and LDL. Various services use the data model to process measurements by linking points, variables, and storage locations.

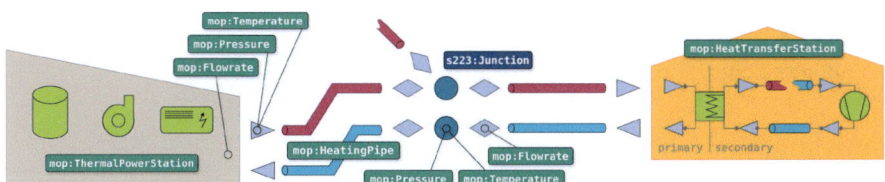

Fig. 3 Principle model of a DHN modelled with some classes from the MOP extension of the 223P ontology

6 Conclusion and Outlook

This paper presented an extension of the 223P ontology for the DHC sector, validated through real use cases in two distinct research projects. The findings demonstrate several advantages in data handling and the development of interfaces to software applications, particularly in terms of machine readability, self-explanatory structures, and standardized data models. The developed extension follows a bottom-up approach, indicating that certain components may be lacking for other use cases and will require further expansion in the future. Despite the promising results, it is evident that many ontologies suffer from a lack of market distribution, hindering the establishment of a "true" standard. Therefore, a community-driven and open-source approach appears to be the most suitable path forward and should be pursued in future developments. In conclusion, while the extension of the 223P ontology demonstrates significant potential for a standardized ontology within the DHC sector, it remains far from achieving standard status due to its specific use cases and the absence of a cohesive community. This highlights the need for further research and active community building in this domain, as further investigations must be carried out in the use cases to evaluate the approach.

7 Funding

This work was funded by the BMWK (German Federal Ministry of Economic Affairs and Climate Action) and the BMUV (German Federal Ministry for the Environment, Nature Conservation, Nuclear Safety and Consumer Protection) under grant 03EN3074A (SimKI-MOP), 03EN3093B (TrafoWaermeNetz) and 13DR013C (3SmartFW).

8 Authors Contribution

All authors contributed equally to the manuscript.

References

1. Prakash AK et al (2024) Ontologies at work: analyzing information requirements for model predictive control in buildings. In: Proceedings of the 11th ACM international conference on systems for energy-efficient buildings, cities, and transportation. https://doi.org/10.1145/3671127.3698189
2. Balaji B et al (2018) Brick: metadata schema for portable smart building applications. Appl Energy 22. https://doi.org/10.1016/j.apenergy.2018.02.091
3. ASHRAE (2023) Standard 223p. https://github.com/open223

4. García-Castro R et al (2023) The ETSI SAREF ontology for smart applications: a long path of development and evolution. Wiley Ltd., Chap 7. https://doi.org/10.1002/9781119899457.ch7
5. Peng Z, Timoudas TO, Wang Q (2025) Critical review of applications of semantic modeling for scalable deployment of data-driven services in 4-5th generation district heating and cooling systems. https://doi.org/10.2139/ssrn.5163003
6. Hippolyte JL et al (2018) Ontology-driven development of web services to support district energy applications. Autom Constr 86. https://doi.org/10.1016/j.autcon.2017.10.004
7. Lefrançois M (2017) Planned ETSI SAREF extensions based on the W3C&OGC SOSA/SSN-compatible seas ontology patterns. In: Proceedings of workshop on semantic interoperability and standardization in the IoT, SIS-IoT
8. Kukkonen V et al (2022) An ontology to support flow system descriptions from design to operation of buildings. Autom Constr 134. https://doi.org/10.1016/j.autcon.2021.104067
9. Pauen N, Schlütter D (2023) TUBES system ontology. https://w3id.org/tso
10. Pruvost H (2023) SENSE - a semantic composition that conceptualizes building energy systems in their operation phase. https://w3id.org/sense/
11. Förster H et al (2024) Open energy academy. https://github.com/OpenEnergyPlatform/academy
12. Ortega S et al (2019) Making the invisible visible - strategies for visualizing underground infrastructures in immersive environments. ISPRS Int J Geo Inf 8(3):15
13. buildingSMART International (n.d.) IFC 4.3.2.0 development. https://github.com/buildingSMART/IFC4.3.x-development/tree/master. Accessed 14 May 2025
14. AGFW (n.d.) https://www.agfw.de/technik-sicherheit/begriffe/. Accessed 14 May 2025
15. BuildingSMART International (n.d.) buildingsmart data dictionary. https://www.buildingsmart.org/users/services/buildingsmart-data-dictionary/. Accessed 15 May 2025

Open Access This chapter is licensed under the terms of the Creative Commons Attribution 4.0 International License (http://creativecommons.org/licenses/by/4.0/), which permits use, sharing, adaptation, distribution and reproduction in any medium or format, as long as you give appropriate credit to the original author(s) and the source, provide a link to the Creative Commons license and indicate if changes were made.

The images or other third party material in this chapter are included in the chapter's Creative Commons license, unless indicated otherwise in a credit line to the material. If material is not included in the chapter's Creative Commons license and your intended use is not permitted by statutory regulation or exceeds the permitted use, you will need to obtain permission directly from the copyright holder.

Optimizing a Multi-vector Energy System with Geothermal-Powered District Heating

Natalia Kozlowska, Arthur Lefebvre, Julien Jacquemin, and Pierre Dewallef

Abstract This paper investigates optimal configurations for multi-vector energy systems, with a focus on low-temperature district heating networks (DHNs) powered by shallow geothermal energy. The system takes into account natural ground regeneration during heat extraction, which prevents borehole freezing and maintains efficiency. The study examines the balance between centralized and decentralized heat production while ensuring fair thermal energy distribution among buildings. The DHN model accounts for heat losses and inter-building distances, with a piecewise linear method used to represent pipe costs. The optimization framework evaluates both building-scale and district-scale solutions, considering various technologies, ground constraints, and temperature levels. The Renewable Energy Hub Optimizer (REHO) is employed as a basis to determine the optimal design and operation of these energy systems [7]. The case study focuses on a building stock in Belgium, incorporating building-level demands for space heating and electricity.

Keywords Optimization · Energy system · District heating network

1 Introduction

DHNs offer a cost-effective way to deliver low-carbon heat at large scale [1]. However, around 90% of district heat still comes from fossil fuels. Achieving Net Zero Emissions by 2050 requires accelerating the shift to renewable sources, such as bioenergy, solar thermal, geothermal, and large-scale heat pumps, and improving network efficiency [1]. The growing development of energy communities enables households to jointly invest in shared infrastructure for heat and electricity generation, including low-temperature DHNs that require centralized coordination. Combining centralized

---

N. Kozlowska (✉) · J. Jacquemin · P. Dewallef
Université de Liége, Département d'aérospatiale et mécanique, Liége, Belgium
e-mail: natalia.kozlowska@uliege.be

A. Lefebvre · J. Jacquemin
Institute of Mechanics, Materials and Civil Engineering, Université catholique de Louvain, Louvain-la-Neuve, Belgium

and decentralized solutions can enhance system flexibility, resilience and efficiency. Designing such hybrid systems requires tools that integrate economic and technical considerations. This study uses a Mixed-Integer Linear Programming (MILP) model solved with Gurobi, where integer variables represent key decisions such as DHN deployment, technology selection, and nonlinear pipe cost modeling.

Various studies have addressed the optimization of district heating networks using MILP approaches, with differing scopes and limitations. For example, centralized system design and operation was focused on in [2] but excluded geothermal heat pumps and decentralized units. Another study detailed in [3] explored distributed energy systems using MILP but did not emphasize technology modeling. Several tools have been developed to support DHN design. Integrated optimization framework was proposed in [4] that includes decentralized units, though limited to heat pumps and storage, but excludes electricity and gas vectors. Similarly, MODESTO was introduced for heat configuration analysis, but did not account for the electricity vector [5]. The nPro tool extended this by incorporating multiple technologies and the electricity vector [6].

REHO offers multi-scale modeling from building to district level, supports various energy vectors, and enables centralized/decentralized configurations [7]. It introduces temperature-based selection via the heat cascade concept, but lacks spatially detailed DHN modeling, such as connection distances, heat losses, nonlinear cost functions, and pump consumption.

This study extends the REHO framework to address key gaps in district heating network design, integrate decentralized and centralized systems, incorporate geothermal heat pumps, and account for ground regeneration. An optimization tool is proposed for small-scale energy systems (up to a few dozen buildings), coupling electricity and heat vectors to capture interdependencies and constraints across both domains. The model also includes battery and thermal storage to support cost-effective and low-fossil energy systems. It enables rapid assessment of system costs, design, and operation with minimal input.

2 Problem Statement

The developed optimization tool is applied to a case study of a building stock in Belgium comprising 17 buildings with space heating and electricity demands. Objectives of the study are to determine an economically optimal mix for space heating and electricity production and distribution and decarbonize the energy system. Domestic hot water and cooling are excluded due to their minor contributions and will be addressed in future work. Heating supply options include centralized solutions, such as a 4th generation district heating network powered by a shallow, closed-loop geothermal heat pump with a field of boreholes, a gas boiler, and thermal storage, and decentralized systems located within buildings, including air-source heat pumps, gas boilers, and electric heaters. The optimization balances these approaches to minimize annual costs.

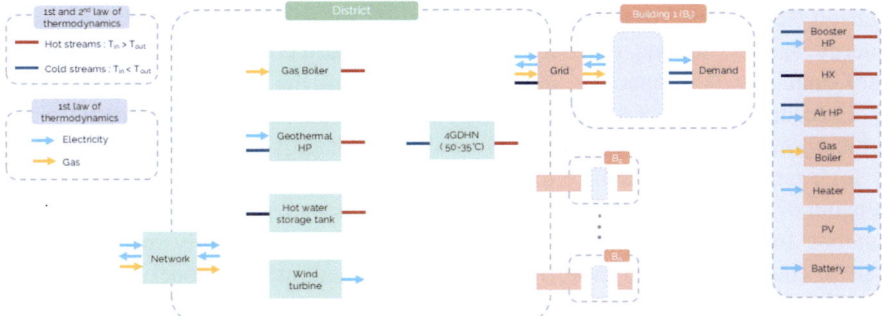

Fig. 1 Multi-vector energy system: individual technologies in orange and centralized technologies in green with connection to the buildings by the grid

Heat delivery from the DHN to buildings is provided via either heat exchangers or booster heat pumps, depending on temperature requirements. The geothermal system accounts for seasonal ground temperature variations, considering both heat extraction and natural regeneration. Figure 1 illustrates an energy system where buildings may meet their energy needs through individual systems, grid-connected electricity exchange, DHN heat, or a hybrid of these options. Electricity demand can eventually be satisfied by PV panels, a wind turbine, a connection to a battery and electricity import.

2.1 Problem Formulation

The optimization problem aims to minimize the total annual cost, consisting of capital (CAPEX) and operational expenditures (OPEX) associated with each building's technologies, centralized technologies, and primary energy use:

$$\min \text{Obj} = \text{CAPEX} + \text{OPEX} \tag{1}$$

To ensure computational tractability, the analysis is based on a set of 10 representative days selected from the full year using a k-medoid clustering approach [9]. Each representative day corresponds to a cluster of similar days characterized by comparable hourly profiles of building energy demand, solar irradiance, outdoor air temperature, wind speed, and electricity and gas prices.

The model incorporates a range of technologies, for both heat and electricity generation, as well as energy storage. These technologies are subject to their own technical constraints, but also capacity limits and energy balance requirements, including thermal and electrical coupling. A comprehensive overview of these constraints is provided in [8]. Investment and maintenance cost data for the decentralized and centralized technologies were found in the EU Joint Research Centre[1] and the Danish Energy Agency[2] reports. Fixed and variable investment costs are used to discourage very small installed capacities and better approximate the cost functions.

To accurately capture long-term energy shifting capabilities, interseasonal storage is incorporated into the model. This allows surplus energy generated during periods of high renewable availability to be stored and later used during periods of high demand or low generation. The model achieves this by reconstructing the full-year timeline from the set of typical days, enabling sequential hourly tracking of the state of charge across the entire year. This approach allows the model to capture the temporal flexibility offered by large-scale thermal storage systems, which is essential for evaluating seasonal energy balancing strategies.

In shallow, closed-loop geothermal heat pump systems with borehole (BH) fields, ground temperatures around each borehole must remain above 0 °C to avoid freezing and ensure efficiency. Each 100 m-deep BH has a radius of 0.068 m, with U-pipe outer and inner radii of 0.016 and 0.013 m, respectively. Typical thermal conductivity values were used [10]. An analytical solution was applied to determine the temperature distribution, yielding a maximum extractable power of 4.23 kW per borehole under the 0 °C constraint [10]. BH interactions and seasonal ground temperature variations were neglected as a first approximation.

2.2 District Heating Network

The district heating network is modeled to account for both heat losses and distance-based household connections. The network topology is predefined, while the decisions to construct the DHN and determine the pipe capacities are optimized based on the trade-off between centralized and decentralized solutions. Energy balance equations are applied at both the arcs and nodes of the network, enabling the modeling of any type of network topology. Each node in the DHN is categorized as one of the following:

- connected to production and/or storage units,
- connected to building heat demand,
- enabling energy flow without associated production or demand.

The energy balance on each node n, and for every time and period, is:

$$\forall n, \forall p, \forall t \sum_{i,j \in A_n} \left(P_{i,j,p,t}^{in} - P_{i,j,p,t}^{out} \right) = \dot{Q}_{n,p,t}^{heat} + \dot{Q}_{n,p,t}^{ch} - \dot{Q}_{n,p,t}^{disch} - \dot{Q}_{n,p,t}^{dem} \quad (2)$$

---

[1] EU Research Centre Report: https://data.europa.eu/doi/10.2760/110433.
[2] Danish Energy Agency, Technology Catalogues: https://ens.dk/en.

where $P^{in}_{i,j,p,t}$ and $P^{out}_{i,j,p,t}$ represent thermal power entering and leaving pipe (i, j) at node n, $\dot{Q}^{heat}_{n,p,t}$ is the heat produced, $\dot{Q}^{ch}_{n,p,t}$ and $\dot{Q}^{disch}_{n,p,t}$ are storage charging and discharging powers, and $\dot{Q}^{dem}_{n,p,t}$ is the heating demand.

The arcs connecting each node, which represent the pipes, are subject to thermal losses along their length. These heat losses are a significant drawback in district heating networks and must be accounted for to ensure accurate sizing. Manufacturer data express heat losses [kW/m] as a function of pipe diameter, following a quadratic trend. To avoid introducing nonlinearities into the model, heat losses are instead approximated as a function of nominal pipe capacity (Fig. 2a) by assuming a constant flow velocity of 1 m/s. This allows the use of the relation $P^{nom} = \rho v c_p A \Delta T$ where P is the nominal thermal capacity. Two linear regressions were then performed on this transformed data for a DHN with a supply temperature of 50 °C and a return temperature of 35 °C. The reference heat loss values [kW/m] are given for a temperature difference of 65 °C, corresponding to an 85 °C/55 °C network and a 5 °C ambient temperature. In our model, the actual ΔT is dynamically calculated as $\frac{T_{supply} + T_{return}}{2} - T_{amb,p,t}$, where ambient temperature varies hourly based on meteorological data. Given that the pipe capacities remain below 2 MW in the studied cases, only the first regression is applied to reduce computational complexity.

The corresponding energy balance on arcs is expressed by the following equation:

$$\forall i, j \in A, \forall p, \forall t \quad P^{out}_{i,j,p,t} = P^{in}_{i,j,p,t} - Q^{loss}_{i,j,p,t} \times L_{i,j} \qquad (3)$$

The cost of DHN pipes generally increases with pipe diameter, but this relationship is inherently nonlinear. Since cost is a critical driver in the optimal design of DHNs, accurately capturing this nonlinear cost behavior is essential. To preserve tractability within the optimization model while approximating the nonlinearity, a piecewise linear (PWL) cost formulation is employed. This approach discretizes the capacity domain into a set of predefined intervals. For each arc of the network, binary variables ($Y^{int}_{i,j,k}$) are introduced to activate one and only one interval, corresponding to the selected pipe capacity. Continuous variables then define the actual capacity ($P^{nom,int}_{i,j,k}$) within the bounds of the selected interval. The total pipe cost is thus modeled as a PWL function of the installed capacity, as shown in Fig. 2b, maintaining a linear structure compatible with MILP solvers. The formulation is expressed by the following constraints:

$$\sum_{k=1}^{n_{cost,int}} P^{nom,int}_{i,j,k} = P^{nom}_{i,j} \quad \forall (i, j) \in \mathcal{A} \qquad (4)$$

$$Y^{built} \geq \sum_{k=1}^{n_{cost,pipe}} Y^{int}_{i,j,k} \quad \forall (i, j) \in \mathcal{A} \qquad (5)$$

$$P^{nom,int}_{i,j,k} \leq c^{max,cost}_k \times Y^{int}_{i,j,k} \quad \forall (i, j) \in \mathcal{A}, \forall k \in \{1, \ldots, n^{cost}_{pipe}\} \qquad (6)$$

$$P^{nom,int}_{i,j,k} \geq c^{min,cost}_k \times Y^{int}_{i,j,k} \quad \forall (i, j) \in \mathcal{A}, \forall k \in \{1, \ldots, n^{cost}_{pipe}\} \qquad (7)$$

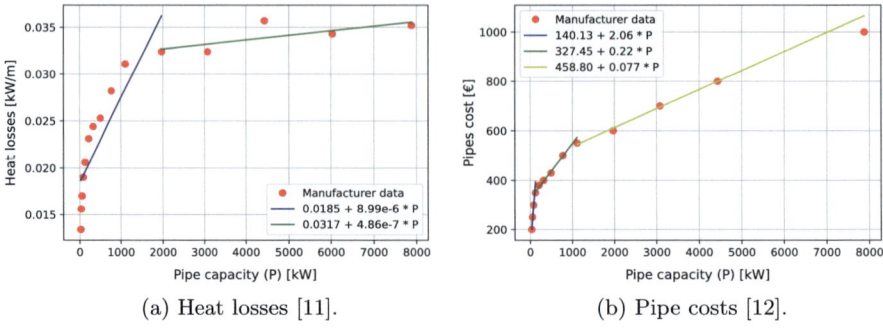

(a) Heat losses [11].   (b) Pipe costs [12].

Fig. 2 Linear regression for PWL programming as a function of pipe capacity

$$\text{Cost}_{i,j} = \tau^{\text{pipe}} \times L_{i,j} \times \sum_{k=1}^{n_{\text{cost,pipe}}} \left(\alpha_k \times Y_{i,j,k}^{\text{int}} + \beta_k \times P_{i,j,k}^{\text{nom,int}} \right) \quad \forall (i,j) \in \mathcal{A} \quad (8)$$

where $c_k^{\text{min,cost}}$ and $c_k^{\text{max,cost}}$ are lower and upper pipe capacity bounds of each interval, $L_{i,j}$ is the length of each arc (pipe), τ^{pipe} is the annuity factor for pipe construction, α_k and β_k are the intercept and slope of the linear equation.

The electricity consumption of the circulation pump in the district heating network is considered by calculating the mass flow rate (9), assuming a constant pressure drop of 100Pa per meter of pipe, a pump efficiency of 75% and a critical path of length L:

$$\dot{m}_{i,j,p,t} = \frac{P_{i,j,p,t}^{\text{in}}}{c_p \times \left(T^{\text{supply}} - T^{\text{return}} \right)} \quad \forall (i,j) \in \mathcal{A}, \; \forall p \in \mathcal{P}, \; \forall t \in \mathcal{T}_p \quad (9)$$

$$\text{Elec}_{p,t}^{\text{dhn}} = \sum_{i \in \mathcal{N}^{\text{prod}}, \, (i,j) \in \mathcal{A}} \dot{m}_{i,j,p,t} \times \frac{\Delta P \times L}{\rho \times \eta_{\text{pp}}} \quad \forall p \in \mathcal{P}, \; \forall t \in \mathcal{T}_p \quad (10)$$

2.3 Case Study

The case study evaluates a single optimization scenario that minimizes annual total system cost to determine the optimal energy mix. The analysis balances centralized and decentralized technologies to identify the most cost-effective configuration. If economically viable, a low-temperature district heating network (50 °C–30 °C) would require 7.8 km of piping (supply and return) to connect all buildings, with centralized heat production located at a single node. The study uses electricity pricing and weather data for the Liége region. The primary goal is to present a design tool needing minimal input to help decision-makers develop integrated energy systems for heating, domestic hot water, electricity, and cooling, emphasizing the economic viability of electrification, particularly via geothermal heat pumps.

3 Results

The optimal energy mix in the reference scenario combines centralized and decentralized technologies. A DHN is selected, with heat exchangers installed in substations to transfer heat to individual buildings.

On the centralized side, a 1019 kW geothermal heat pump is implemented, coupled with a thermal storage tank of 344kWh. The system includes 241 boreholes, each 100 m deep and spaced 8 m apart, covering a geothermal field of 13440 m^2. With an 8 m borehole spacing, thermal interference is negligible, and ground temperature remains above 0 °C, confirming sufficient natural regeneration. As shown in Fig. 3, the heat pump charges the storage tank during off-peak hours in low-demand periods, which is then discharged during peak hours. On low-demand days, storage is mainly charged during daylight using PV electricity, compensating for night-time heat losses in the DHN. Despite the presence of a central plant, a significant portion of demand is still covered by decentralized systems.

Installed capacities per building are shown in Fig. 4. Gas boilers and air-source heat pumps are selected in nearly all buildings with decentralized systems, often operating in parallel with the DHN through heat exchangers. Two buildings (B4 and B8) include electric resistance heaters as backup for peak loads. However, the high number of technologies installed in some buildings, particularly B4 and B8, suggests future work such as using a piecewise linear formulation for investment costs to better account for economies of scale, or performing sensitivity analyses on demand and primary energy prices to improve the realism of the design.

The annual total system cost for the reference scenario is 2085 k €(648 k€for CAPEX, 1259 k€for OPEX, 58 k€for piping costs). Table 1 summarizes the district heating network performance, showing moderate heat losses (13.3%) and efficient pump operation with 79MWh of annual electricity use.

Total electricity demand, including both building and heating technologies, is partly covered by a 2 MW wind turbine and 4778 kW of PV panels. Remaining

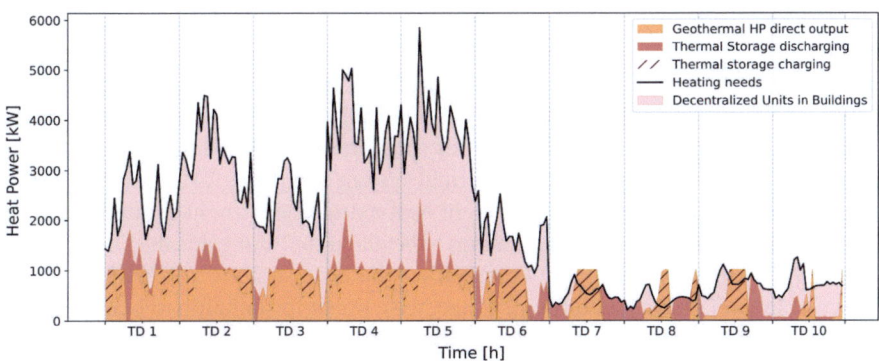

Fig. 3 Load curve of centralized and decentralized technologies

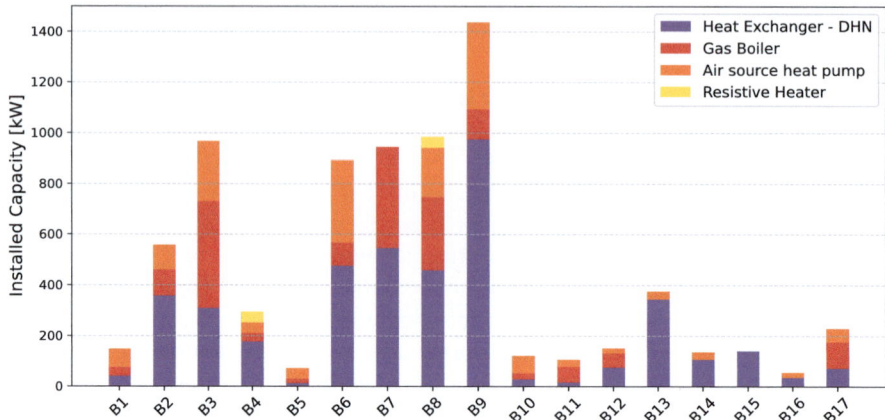

Fig. 4 Installed decentralized technologies in every building

Table 1 District heating network indicators

Indicator	Value
Annual DHN thermal energy [MWh$_{th}$]	4633.85
Annual heat losses [MWh$_{th}$]	615.16
Share of heat losses [%]	13.28
Linear heat density [MWh/m/an]	0.58
Pump consumption [MWh$_{el}$]	78.99

demand (6274 MWh/year) is met by grid imports. Due to renewable intermittency, some locally generated electricity is exported back to the grid.

4 Conclusion

This work proposes a methodology for the preliminary design of multi-vector energy system, enabling the identification of an economically optimal energy mix combining centralized and decentralized technologies, supported by a low-temperature district heating network (50–35 °C). The geothermal system comprises 241 boreholes 100 m deep spaced 8 m apart, ensuring sufficient heat extraction without ground freezing. Significant decentralized capacity remains to be integrated alongside the DHN, which shows 13.28% thermal losses and a circulation pump consumption of 79 MWh. DHN pipe costs are approximated through a piecewise linear formulation to better capture their nonlinear behavior. The proposed tool enables early-stage design and operation with minimal input data. Future works will address cooling needs and artificial ground regeneration integration and the inclusion of electrical network modeling to capture combined thermal and electrical design constraints.

References

1. International Energy Agency (2023) District heating. https://www.iea.org/energysystem/buildings/district-heating
2. Morvaj B, Evins R, Carmeliet J (2016) Optimising urban energy systems: simultaneous system sizing, operation and district heating network layout. Energy 116:619–636
3. Omu AC, Boies R, Adam (2013) Distributed energy resource system optimisation using mixed integer linear programming. Energy Policy 61:249–266. https://doi.org/10.1016/j.enpol.2013.05.009
4. Rojer J, Janssen F, van der Klauw T, van Rooyen J (2024) Integral techno-economic design and operational optimization for district heating networks with a mixed integer linear programming strategy. Energy 308
5. Vandermeulen A, van der Heijde B, Vanhoudt D, Salenbien R, Helsen L (2018) Modesto - a multi-objective district energy systems toolbox for optimization
6. Wirtz M (2023) nPro: a web-based planning tool for designing district energy systems and thermal networks. Energy 268
7. Lepour D, Loustau J, Terrier C, Maréchal F (2024) REHO: a decision support tool for renewable energy communities. J Open Source Softw 9(103):6734
8. Jacquemin J, Kozlowska N, Lefebvre A, Jeanmart H, Quoilin S (2025) Impact of energy sharing on households' energy system design and management, under specific electricity cost structures. ECOS, Paris, France
9. Limpens G, Moret S, Jeanmart H, Maréchal F (2019) Energyscope TD: a novel open-source model for regional energy systems. Appl Energy 255:113729
10. Coen T, François B, Gerard G (2021) Analytical solution for multi-borehole heat exchangers field including discontinuous and heterogeneous heat loads. Energy Build 253:111520. ISSN 0378-7788
11. Products catalog, PowerPipe, Sweden. https://www.powerpipe.se/en/le-catalogue-des-produits
12. Resimont T (22 Dec 2021) Strategic outline and sizing of district heating networks using a geographic information system. PhD thesis, ULiege

Open Access This chapter is licensed under the terms of the Creative Commons Attribution 4.0 International License (http://creativecommons.org/licenses/by/4.0/), which permits use, sharing, adaptation, distribution and reproduction in any medium or format, as long as you give appropriate credit to the original author(s) and the source, provide a link to the Creative Commons license and indicate if changes were made.

The images or other third party material in this chapter are included in the chapter's Creative Commons license, unless indicated otherwise in a credit line to the material. If material is not included in the chapter's Creative Commons license and your intended use is not permitted by statutory regulation or exceeds the permitted use, you will need to obtain permission directly from the copyright holder.

Design Optimization of District Heating Networks with Thermal-Hydraulic Validation

Yacine Gaoua, Charlie Pretot, and Nicolas Vasset

Abstract District heating networks (DHNs) are key to energy transition strategies, offering cost-efficient, reliable and sustainable thermal energy distribution. However, obtaining realistic and efficient designs in dense urban environments is a demanding task. Critical factors like optional user connections, field constraints, multiple production scenarios, and thermal-hydraulic viability are often overlooked in typical engineering approaches. To overcome these limitations, a planning tool integrating graph theory, mathematical optimization, and thermal-hydraulic simulations was developed. The problem is solved through selecting consumers from prospects and computing optimal layouts, pipe diameters, and consolidated costs. Flow velocity regimes, demand diversity and temperature regimes at substations are included. A thermal-hydraulic simulator concurrently validates performance for considered scenarios. A realistic design case based on available data is presented, exhibiting in particular the impact of the chosen maximisation objective on results.

Keywords District heating networks · Mathematical optimization · Thermal-hydraulic simulation

1 Introduction

The development of new (greenfield) district heating networks, or geographical extensions of existing ones, is currently encouraged in institutional roadmaps (see for example [1]). Such long-term infrastructure planning requires careful evaluation by future conceding authorities, as it involves long-lasting economical and technical decisions. As architectures also evolve towards 4th generation concepts [2], A variety of production sources must be considered, with possible partial availability or reliability. Several heat provision scenarios must then be considered, as it impacts the network design at first order. The design must also include decisions concerning

Y. Gaoua · C. Pretot · N. Vasset (✉)
University Grenoble Alpes, CEA, LITEN, INES, Le Bourget-du-Lac, France
e-mail: nicolas.vasset@cea.fr

which prospects to include, and which to discard. Finally, operational constraints must be mitigated against whilst minimizing investment costs.

Advanced design strategies exist in scientific literature, including direct application of graph theory [3], linearized approaches or heuristics [4]. As one of the main concerns of such approaches is representativity, it is indeed difficult to guarantee in those cases that the obtained design will be operationally viable beyond *a priori* hypotheses. DHC design initiatives that include substantial physical modeling take either the route of metaheuristics supported by a simulator [5], or direct formulation of an optimization problem, leading to non-linear optimizations [6]. Among the latter works, [7] allows to tackle reasonably large problems with a multi-scenario approach. A current limitation appears to be the flexibility on the list of prospects for the solution sought, this one being *a priori* fixed before optimization takes place.

In order to try and leverage the different limitations described above, a methodology supported by a tool, and intended for design engineers was developed. For this, we developed a framework converting raw geographical data (public roads, critical infrastructure and buildings) to a graph-type problem with parameterized physical inputs for thermal operation. This problem is solved through Mixed Integer Linear Programming (MILP) optimization, hypothesis parameters being then refreshed iteratively using a thermal-hydraulic simulator [8], until convergence. The approach allows also to take into account multiple scenarios and remain fully flexible for the selection of network consumers.

2 Methodology

2.1 Formulation

2.1.1 Notations and Graph Basic Formulation

The public road layout is reduced to a set of non-oriented edges and nodes $\{E_{ij}, N_i\}$. Other subsets of nodes N_i^c, N_i^p are possible consumer connections and production sites. A network design consists in selecting items $E_{ij}^\star$ and $N_i^\star$. Note that in this formulation, an edge will implicitly carry a supply and a return line.

2.1.2 Graph Reduction

Heuristics are put into place in the original graph in order to reduce problem cardinality. On a first level, all nodes with a connectivity of exactly 2 are removed, and adjacent edges are replaced by a single edge with length the sum of the replaced ones. On a second level, the original problem graph is reduced by only conserving edges appearing in outlines of shortest paths between any two nodes in the subset $\{N_i^c, N_i^p\}$. This manipulation substantially lowers the number of edges to consider while conserving the ability to find an optimal layout in the remaining edges, with no observed degradation.

2.1.3 Physical Hypotheses

We try to solve network operation for a set of scenarios $\{\xi_k\}$, defined by characteristics in production and consumption. In particular, for each production site, a maximal thermal power output $\overline{\mathcal{P}_i^p}$ is defined, and a scenario is materialized by a relative maximal engagement $\overline{\alpha}_{\xi_k,i}$:

$$\forall i,\ \forall \xi_k,\ \mathcal{P}^p_{i,\xi_k} \leq \overline{\alpha}_{\xi_k,i}\overline{\mathcal{P}_i^p}\ ;\ \overline{\alpha}_{\xi_k,i} \leq 1. \tag{1}$$

Actual engagement $\alpha_{\xi_k,i}$ verifies $\alpha_{\xi_k,i} \leq \overline{\alpha}_{\xi_k,i}$. The governing equation for network operation is the continuity equation throughout the whole network, written for each scenario as balance equations at each node:

$$\sum_{j\in\mathcal{N}_{\text{in}}(i)} \dot{m}_{ji} - \sum_{k\in\mathcal{N}_{\text{out}}(i)} \dot{m}_{ik} = -\dot{m}_i^c + \dot{m}_i^p, \tag{2}$$

where $\mathcal{N}_{\text{in}}(i)$ and $\mathcal{N}_{\text{out}}(i)$ are the set of nodes with edges entering (resp. leaving) node i, $\dot{m}_{ji}$ is the mass flow modeled through edge E_{ji}, algebraically from N_j to N_i, and $\dot{m}_i^c$, $\dot{m}_i^p$, if node N_i is a consumer (resp. producer), are the mass flow rates flowing at that point from the supply to return (resp. return to supply) line (zero by default). All values are given for a specific scenario. At producer or consumer nodes, heat provision (resp. delivery) is then materialized by the link between a requested power $P_i^{p/c}$ and mass flow $\dot{m}^{c/p}$ through the water heat capacity C_p and upstream/downstream temperature difference $\Delta T_i^{p/c}$:

$$\mathcal{P}_i^{p/c} = \dot{m}_i^{c/p} C_p \Delta T_i^{p/c} \tag{3}$$

Finally, on the optimization problem, a thermal loss factor $\theta_{loss} \in [0, 1[$ is defined by default so that it represents global loss in the network in the form of thermal power. It is used to link temperature differences at the production and distribution sides, so that this factor effectively models homogeneous power loss in the network; all values for temperature differences in production units being considered equal in this modeling approach.

$$\forall i,\ \forall j,\ \Delta T_i^c = \theta_{loss} \Delta T_j^p \tag{4}$$

This thermal loss computation, alongside ΔT values will be overridden when iterating on hypotheses with the thermal-hydraulic simulation.

2.1.4 Economical Hypotheses

Implementation costs include four components: Production plants can be assigned a cost of installation c_i^p, labeled in €/MW. Consumers are assigned a cost with two dependencies: a power-dependent cost for substation installation c_i^c, in €/MW, and a cost related to the building/public road connection with appropriate length and dimensioning (see later). We use for pipe installation costs a typical catalogue of values in €/m for diameters ranging from 50 to 500mm. Finally, on the original map, the layout of several critical paths in a geographical zone is constructed and labeled as a 1D discontinuous feature Λ. A fixed cost c_Λ is added to a global budget for every selected edge that crosses this critical path. This would correspond to real-life geographical constraints which typically orientate engineers towards mitigated solutions.

2.1.5 Thermal-Hydraulic Modeling and Physical Parameter Update

The formulation for optimization will only include parameterized equations for thermal behavior, the only first-principles equations enforced being the continuity equations. Those arbitrary parameters are to be updated through thermal-hydraulic simulation, which we outline here.

A global description of the convergence scheme is described in Fig. 1. From a first optimization resolution, a network layout and dimensioning will be modeled using a dedicated thermal hydraulic simulator [8]. This simulator requests additional information such as production plant setpoint temperature $T_{i,set}^p$, temperature regime on the secondary side and physical models such as pipe model with insulant description. We refer the reader to the above reference for the description of thermal-hydraulic model. With multiple power sources, the simulator is tuned using pressure difference setpoints on each hydraulic pump situated in production plants, so that plant engagement ratio $\alpha_{\xi_k,i}\overline{\mathcal{P}_i^p}/(\alpha_{\xi_k,j}\overline{\mathcal{P}_j^p})$ is respected. In the end, a flow configuration is found in the network, including a discretized estimation of thermal and hydraulic losses, thermal and hydraulic regime in the primary side for all substations, and return temperature at production plants. Those outputs are used to update parameters $\Delta T_i^c, \Delta T_j^p$ for each i, j, and each scenario ξ_k, here with a simplified notation. The optimization is then launched again. With new values for engagements $\alpha_{\xi_k,i}$ and a

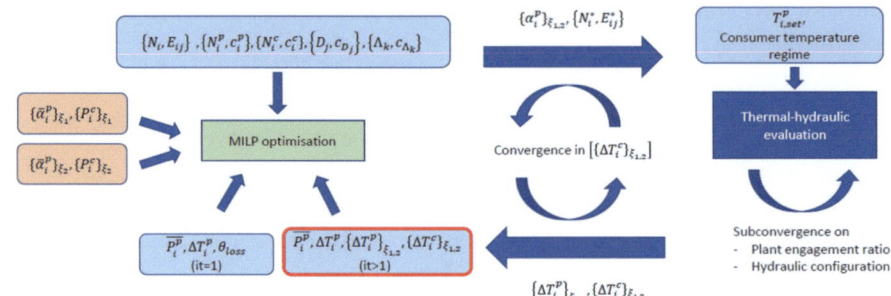

Fig. 1 Description of the convergence principle for the coupling between the optimizer and the thermal-hydraulic simulator

new global layout, another iteration can be performed. Convergence is attained on a stopping criterion based on the maximal variation of ΔT_i^c between two steps.

2.1.6 Additional Heuristics and Objectives

In addition to physical feasibility, the main constraint imposed to a potential design is a velocity constraint. We impose all velocities in the network to be below the value of 2 m/s for every scenario, this value being computed using an assumed hydraulic diameter being equal to the geometrical diameter of pipes. In order to take into account demand diversity, an additional post-computation heuristic is used for enforcing this velocity constraint: notwithstanding the computed mass flow rate values in the converged solution, and for every scenario, the velocity constraint will be only imposed assuming only a fraction η of the mass flow rate depending on the pipe. This is done by defining a value η_D for all pipes where mass flow rate in all scenarios contributes to the provision of at least D consumers. Typical values used in the remainder of this work are $\eta_D = 1$ for $D < 3$, $\eta_D = 0.9$ for $D \in [\![3, 10[\![$ and $\eta_D = 0.8$ for $D \geq 10$.

There is a practical need to associate building locations where a potential consumer is located, to one or several connection points in the public road. We developed a heuristic to systematize this connection step, with a computed pipe route taking into account building shape and choosing the closest route available. Dimensioning of this layout is computed and added to consumer connection costs.

Finally, formal objectives for the optimization are of three different types: an objective in energy is defined, with the optimizer trying to connect the maximal number of prospects available. A penalty term on network length ensures in this case non-degeneracy of the solution. A second option is to label the objective in MW h/m, as the total energy provided in one year with respect to piping length. Finally, a third formulation targets an economic value (relative CAPEX investment) in €/kW, the power value corresponding to the maximal demand level. All three objectives are linearized using classical techniques.

2.2 *Implementation*

The optimisation problem is formalized using the GAMS modeler [9]. This modeler appears very useful for formulating network-type mathematical optimization problems, at it combines multi-dimensional algebraic notation with the concept of relational model. The problem is formulated as a MILP-type optimization problem, which is made possible through our parametric approach for temperature regimes and thermal losses, further corrected through concurrent thermal-hydraulic simulation. We use IBM CPLEX solver [10] for pure resolution of the optimization side.

Problem setup, execution control and coupling with the thermal-hydraulic simulator is performed in Python. In particular, coupling with the DistrictLab Simulation Studio software is performed using FMU encapsulation of the simulator [11].

3 Case Study

3.1 Input Data

A support case for this work is one of a middle-sized city in southeast France. The situation is one of a greenfield project, and is shown in Fig. 2. Two production sites are identified; Site A (East) will be considered as fully reliable and pilotable (a CHP plant, for example), with a maximal power value of 25 MW; Site B (West) is a non-pilotable source (waste heat recovery from an industry, for example) of 15 MW at most, supplemented with a backup production source of 10MW. We do not include the installation cost of the production units in the problem. Prospects are identified using the institutional ENREZO open data source [12], conserving only consumers of more than 300 MW h per year. An arbitrary factor of 1/1000 is considered between annual consumption in MW h and peak consumption power in MW, a value used

Fig. 2 The different zones for prospective targets for the city of interest: Zone 1 (light blue) is South, Zone 2 (green) is East, zone 3 (light orange) is North. Hard solid lines materialize critical sections with additional crossing costs (railroad excluding a viaduct section, river). All represented buildings are prospective users of the network. The green dots correspond to the two production sites (A et B)

in all scenarios under study here. Three differents zones where prospects are looked for are arbitrarily defined, allowing to obtain different design problems of increasing size. Heat exchanger installation costs are fixed at a uniform, arbitrary value. Critical sections (railroad, river) with additional crossing costs (typically 500000€ per crossing) are defined. Public road layout, as well as buildings layout are extracted from the IGN public database BDTOPO [13]. Finally, pipe installation costs are also arbitrarily defined with a price per meter (incl construction works) directly related to the pipe diameter with a quadratic trend. Twelve diameter values from DN50 to DN500 are allowed, with a maximal price ratio of 5.5.

3.2 Description of the Optimization Cases Performed

A first set of computations (Batch 1) are presented as single-production cases with only site A available. Results are presented with objectives on Energy, Energy density and CAPEX per installed power. In those computations, the optimization only requires one scenario, corresponding to full demand satisfaction by Site A. Cases consider increasing perimeters for potential consumers. A second set of computations (Batch 2) contains two-site productions cases with the full perimeter including all three zones. Two dimensioning scenarios are defined, with maximal power values $\{\overline{\mathcal{P}^p}_{A/B}\}_{\xi_{1/2}} = \{(10, 25), (25, 10)\}$ MW. Default hypotheses on first iterations are $\theta_{loss} = 0.95$, $\Delta T_i^p = 40\,°C$. We assume additionally a production temperature of 105 °C and a uniform secondary temperature regime for consumers of [90, 60] °C. All results are obtained under 20 min wallclock time on an Intel(R) Core(TM) Ultra 7 155H 1.40 GHz laptop, with at most four iterations for convergence between the optimisation and the TH simulator.

4 Results

Table 1 (higher panel) sums up results obtained in Batch 1 runs. On Energy density and CAPEX objectives, additional constraints on minimal installed power for the prospective networks are imposed, respectively at 5MW, 10MW and 20MW for the three perimeters.

An interesting trade-off is observed when comparing density objective runs and €/kW-type objective runs, as a decrease of 12% in costs per unit power is observed for the perimeter of Zones 1–2. This trade-off, which is systematically observed, is obtained partially because of the capacity of runs to select appropriate prospects, and then potentially diminish additional costs related to crossing of critical sections. This is clearly a first-order information for a design engineer in order to make informed choices on this respect.

Table 1 Results for Batch1 and Batch 2 runs. Objectives on Energy (E), Energy density (D) and Capex (C) are detailed for perimeters including an increasing number of zones for Batch 1, and for the whole perimeter on Batch 2. Installed power is the total power extracted from the network from selected consumers. Results on installation costs are expressed on a relative basis, with reference being the energy optimization case for each set of runs (same zones, same production hypothesis). Length and density values are computed based on the public road layout. On Batch 2 runs, site engagements are shown for scenario 1 and scenario 2 respectively.

Batch1 runs									
Parameter	Zone1			Zone1-2			Zone1-2-3		
Max nb. of consumers	20			30			61		
Objective	E	D	C	E	D	C	E	D	C
Installed power (MW)	8.72	7.54	7.85	15.22	10.36	12.52	23.75	20	20.51
Network length (m)	3301	2343	2611	5556	2568	3412	8883	5185	5875
Energy density (MWh/m)	2.64	3.22	3	2.74	4.04	3.67	2.67	3.86	3.49
Relative installation cost per kW (-)	1.0	0.98	0.84	1.0	0.90	0.79	1.0	0.83	0.81

Batch2 runs			
Objective	E	D	C
Installed Power (MW)	29.13	20.03	24.75
Network length (m)	11181	5776	7962
Energy density (MWh/m)	2.61	3.47	3.11
Relative installation cost per kW (-)	1	0.903	0.801
Site A engagement (MW)	[25, 4.39]	[20.16, 10]	[14.94, 10]
Site B engagement (MW)	[4.39, 25]	[0, 10.17]	[9.89, 14.89]

Table 1, lower panel sums up results obtained in Batch 2 runs, where a minimal installed power of 20MW is always requested on a perimeter including all three defined zones. In addition, Fig. 3 presents layouts obtained for Batch 2 runs with objectives in energy density and investment costs.

Analysis is similar to Batch 1 on objectives, although very different architectures are obtained. A clear tradeoff between energy density and CAPEX value is also obtained. No consideration of operational costs in the objectives is made here, which explains why potentially cheaper energy sources (on Site B) are not favored in production. This is a potential extension of our work.

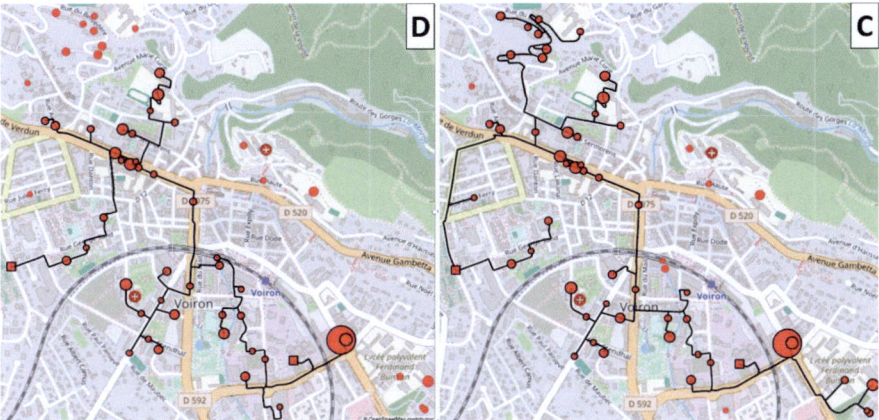

Fig. 3 Obtained layout for the Batch 2 optimisation case with objectives in energy density (D) and investment cost per power unit (C). Red circles correspond to potential prospects with a size dependency on global annual consumption as a visual aid. Black circled ones are selected for network connection. Red square boxes are locations for production

5 Conclusions

We have presented here several aspects of a global design methodology for greenfield projects of district heating networks, or substantial geographical extension of existing ones. One asset of the work presented here is the approach flexibility, allowing for the final solution to include only part of the prospective consumers, depending on sought objectives. At the same time, obtained solutions are systematically evaluated against full-scale thermal-hydraulic (stationary) simulation in all scenarios of interest, to ensure a realistic design. Results show the importance with this problem setup of the optimization target, and the capacity of the approach to handle tradeoffs in this respect.

This work is to be extended in the future to include direct hydraulics resolution inside the optimisation problem, in order to alleviate possible (but still unmet) convergence difficulties. Issues such as network remeshing in order to guarantee operational resilience are also a target for future work. Inclusion of asymmetrical operational costs between production plants could be also a target for future studies.

Acknowledgements This work was initiated with, and partly supported by the Grenoble District Heating Company (CCIAG). The implementation of the tool supporting the methodology presented here is one outcome of a common initiative with our institute (CEA). All data and detailed results for this work are available from the authors upon request for benchmarking/comparison purposes.

References

1. French National low-carbon strategy (2020) www.ecologie.gouv.fr/sites/default/files/documents/2020-03-25_MTES_SNBC2.pdf
2. Lund H, Østergaard PA, Nielsen TB, Werner S, Thorsen JE, Gudmundsson O, Arabkoohsar A, Mathiesen BV (2021) Perspectives on fourth and fifth generation district heating. Energy 227. https://www.sciencedirect.com/science/article/pii/S0360544221007696
3. Kuper L, Metzger M, Stursberg P, Niessen S.: Computationally efficient topology design of district heating networks by price-collecting Steiner trees. SSRN preprint https://doi.org/10.2139/ssrn.5080166
4. Neri M, Guelpa E, Verda V.: Design and connection optimization of a district cooling network: mixed integer programming and heuristic approach. Appl Energy 306. https://www.sciencedirect.com/science/article/pii/S0306261921012939
5. Merlet Y, Baviere R, Vasset N (2023) Optimal retrofit of district heating network to lower temperature levels. Energy 282. https://www.sciencedirect.com/science/article/pii/S0360544223001655
6. Wack Y, Baelmans M, Salenbien R, Blommaert M (2023) Economic topology optimization of district heating networks using a pipe penalization approach. Energy 264. https://www.sciencedirect.com/science/article/pii/S036054422203047X
7. Wack Y, Sollich M, Salenbien R, Diriken J, Baelmans M, Blommaert M (2024) A multi-period topology and design optimization approach for district heating networks. Appl Energy 367. https://www.sciencedirect.com/science/article/pii/S0306261924007633
8. Baviere R, Vallee M, Crevon S, Vasset N, Lamaison N (2023) DISTRICTLAB-H: a new tool to optimize the design and operation of district heating and cooling networks. In: Proceedings of the DHC symposium 2023 - 18th international symposium on district heating and cooling. Beijing, China. https://cea.hal.science/cea-04247969
9. The General Algebraic Modeling Language. https://www.gams.com
10. IBM ILOG CPLEX Optimizer. www.ibm.com/products/ilog-cplex-optimization-studio/cplex-optimizer
11. The Functional Mockup Interface Standard. https://fmi-standard.org
12. CEREMA, ENREZO - French National cartographic heat and cold needs. https://reseaux-chaleur.cerema.fr/cartographie-nationale-besoins-chaleur-froid
13. French Geographical Institute, BDTOPO geographical database. https://geoservices.ign.fr/bdtopo

Open Access This chapter is licensed under the terms of the Creative Commons Attribution 4.0 International License (http://creativecommons.org/licenses/by/4.0/), which permits use, sharing, adaptation, distribution and reproduction in any medium or format, as long as you give appropriate credit to the original author(s) and the source, provide a link to the Creative Commons license and indicate if changes were made.

The images or other third party material in this chapter are included in the chapter's Creative Commons license, unless indicated otherwise in a credit line to the material. If material is not included in the chapter's Creative Commons license and your intended use is not permitted by statutory regulation or exceeds the permitted use, you will need to obtain permission directly from the copyright holder.

Large-Scale Solar Thermal Systems in District Heating Networks: A Review of German Projects Regarding Dimensioning, Temperatures and Stagnation Times

Bert Schiebler, Maik Kirchner, Julian Jensen, and Federico Giovannetti

Abstract The district heating sector in Germany needs to be supplied increasingly from renewable sources. Large-scale solar thermal systems can make a significant contribution to this process. The paper analyzes 38 concepts of feasibility studies as well as 30 realized systems by investigating relevant designing parameters. Network supply temperatures of around 75 °C and 80 °C are very common for the feed-in of solar thermal energy. The achievable solar fraction is strongly dependent on the storage capacity. A doubling of solar fraction typically requires a tenfold increase in storage volume. The installed storage capacity is often smaller than 100 l/m². Large seasonal storages, as usual in Denmark, are very seldom in Germany so far. Thus, stagnation events become more relevant in large-scale systems, as we report from the realized projects. Systems with solar fractions of around 20%, which are typical for district heating networks especially in countryside regions, achieve up to 50 stagnation days per year, a representative stagnation day is analyzed for one monitored system. Finally, the paper addresses possible causes of stagnation and discusses a prevention strategy by using heat pipe collectors with inherent temperature limitation.

Keywords Large-scale solar thermal systems · Stagnation · Heat pipe · Overheating prevention · Cost reduction

1 Introduction

Large-scale solar district heating systems (SDH) has a great potential to decarbonize the heating sector and are becoming increasingly important for achieving climate goals [1]. In parallel to the transformation of district heating sector, the overall share

---

B. Schiebler (✉) · M. Kirchner · J. Jensen · F. Giovannetti
Institute for Solar Energy Research (ISFH), Emmerthal, Germany
e-mail: schiebler@isfh.de; info@isfh.de

of network-based heat supply in Germany is expected to increase significantly from around 14–30% until 2030 [2]. At the same time, the share of renewable energy needs to be significantly increased in the coming years. To supply the still high network temperatures, high-efficiency collectors, such as evacuated tube collectors (ETC), are well suited [1]. Large-scale solar thermal systems are already established to feed district heating networks, whereby a trend towards ever-larger systems can be observed [3, 4].

Based on feasibility studies and realized systems, we investigate current dimensioning practices by analyzing relevant design aspects. We especially examine the solar fractions (SF) in correlation with the number of stagnation events in SDH. Stagnation refers to standstill of the solar circuit under high irradiance, often caused by low summer heat demand. Stagnating collectors can result in overheating and steam formation, which spreads in the piping system at high temperatures. By using specially designed heat pipe collectors, the maximum collector temperature can be limited to around 130 °C. In addition to reducing possible risks and costs associated with stagnation, the intrinsically safe overheating prevention offers optimization potential in planning and implementing solar circuits. This approach has already been successfully demonstrated in small, decentralized solar thermal systems [5]. In the R&D project *HP-BIG*, we are developing such collectors as large-scale modules along with an innovative system concept based on them.

2 Data and Methods

In collaboration with the company *empact GmbH*, we analyzed feasibility studies as part of the German program *Wärmenetze 4.0*. The work includes 14 different locations with different specifications of the heating network, e.g. the individual network temperature requirements. For each location, we considered multiple concepts in the scope of the feasibility studies (2–5 pcs per location), whereby each integrated heat supply concept is based on a solar thermal collector field with different system dimensions. Thus, a total number of 38 system concepts at 14 locations are the basis for our evaluation. Around 70% of the underlying locations represent urban networks with a high share of existing, non-renovated buildings resulting in comparatively high requirements for space heating energy and supply temperatures. Only around 30% of the investigated studies were carried out for quarters with mainly new buildings and correspondingly lower space heating demands. Due to the fact that large parts of the German heat demand still occur at a comparatively high temperature level, the considered feasibility studies provide a representative reflection of the current situation.

We additionally analyzed nearly all realized or ongoing SDH projects in Germany using ETC and exceeding 1000 m^2 in collector area. This sample accounts for about 50% of all German large-scale SDH installations, regardless of collector type. We focused on high-efficiency ETC systems due to the still high temperature requirements of the most district heating networks. Among the 30 analyzed systems, 57%

are located in rural areas (e.g., "bioenergy villages"), while 43% supply urban district heating networks with typically higher supply temperatures. Beyond general design aspects, we also received information on the number of stagnation events directly from the operators. We examined one representative system in detail using measured operational data.

The analysis focuses on system parameters that are particularly relevant to stagnation in large-scale solar thermal systems, including network temperatures, storage volume, solar fraction, and the number of days with at least one stagnation event.

3 Results from Analysis

3.1 Network Temperatures

Figure 1 shows the distribution of the network supply temperatures for the feasibility studies and the realized projects, whereby a differentiation between the summer case and the winter case is usual according to an outdoor temperature dependent operation mode. The winter case represents the condition at very low outdoor temperatures. For SDH systems, the summer case is more relevant due to the much higher irradiation. The relevance of the winter temperatures increases with higher share of annual SF. Based on the analyzed studies, most summer temperatures (57%) are in the range 70–80 °C. The temperature range 60–70 °C is also prominent (29%). The cases ≤ 60 °C and > 90 °C are without further relevance (7% for each). The winter temperature requirements are visibly higher, whereby more than a half still in the range ≤ 80 °C (57%). Based on the realized projects the majority of the summer temperatures are in the range 70–90 °C (87%). In winter, the most SDH are in the range 80–90 °C (60%). Temperatures > 100 °C also represent a relevant share (23%). The median values for the summer supply temperature are 75 °C (feasibility studies) and 80 °C (realized projects), which match very well with results from a previous study of 13 systems in Austria [6].

3.2 SDH Sizing

The area of the collector field and capacity of the thermal energy storage (TES) as well as the corresponding SF are some relevant dimensioning parameters of SDH systems. The SF represents the supplied annual solar yield in relation to the total demand of the heating network. To improve the comparability between smaller and very large systems, we focus on SF and the TES volume in relation to the collector area. Figure 2 shows the SF over the specific TES capacity for 38 study concepts and 24 realized projects (only for data sets with SF > 3%). The data exhibit a significant scattering, which is due to various unconsidered aspects. For example, the TES

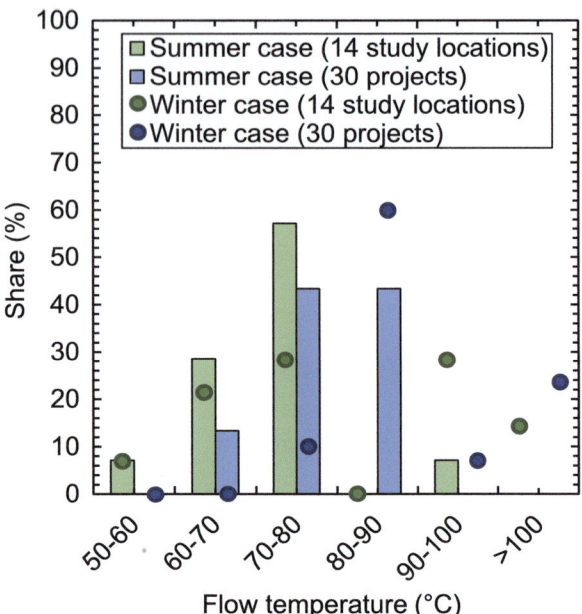

Fig. 1 Distribution of network supply temperatures for 14 feasibility study locations and 30 realized projects, shown for typical summer and winter conditions

capacities are not exclusively used by the collector field, but also by other heat generators in multifunctional operation. Nevertheless, both regressions curves are almost identical and show the SF in dependency on the TES volume. A doubling of the SF from 15 to 30% requires an increase in TES volume by factor 10. The median values of SF and TES capacity for both data sets are in the range between 17 and 20% as well as 90 l/m^2 and 100 l/m^2 respectively. The average for the 48 Austrian systems featured in [6] is 82 l/m^2, which well confirms our results for Germany.

3.3 Amount of Stagnation days

More than 60% of all realized projects have TES < 100 l/m^2, whereby the addressed annual SF is often above 20% (in 36% of all systems). Especially small SDH (e.g. bioenergy villages) are often designed for achieving higher SF, due to the better availability of ground areas and the tendency towards comparatively lower network temperatures. Thus, many plant operators reported that unpleasant start-stop behavior of the biomass boilers can be completely avoided in the summer months and fuel as well as emissions can be saved by using solar thermal systems. The comparatively small TES capacities lead to relevant stagnation events in summer.

Figure 3 shows the number of stagnation days per year plotted in dependency of SF. For systems with SF around 20%, we have determined an average stagnation time of 25 days. The tolerance band (+100% and − 50%) covers almost the complete

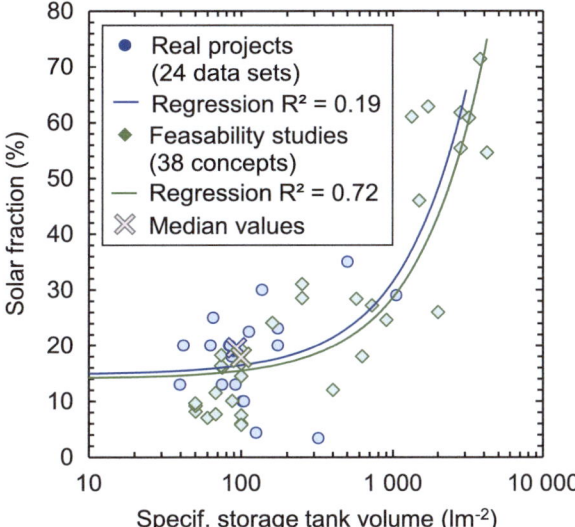

Fig. 2 Solar fraction over specific storage volume for 38 concepts of the feasibility studies and 24 data sets of the realized projects (criteria SF > 3.0%)

data range and illustrate the high variability of the results. The variance results from technical properties (e.g. TES capacity) or from type of data source. Accordingly, 13 to 50 stagnation days are to be expected for SF = 20%. Therefore, this system state must be given an appropriate attention in designing large-scale SDH systems. In practice, the handling of stagnation should be intrinsically safe, for example by draining the collector field [7] or using suitable avoidance strategies [5].

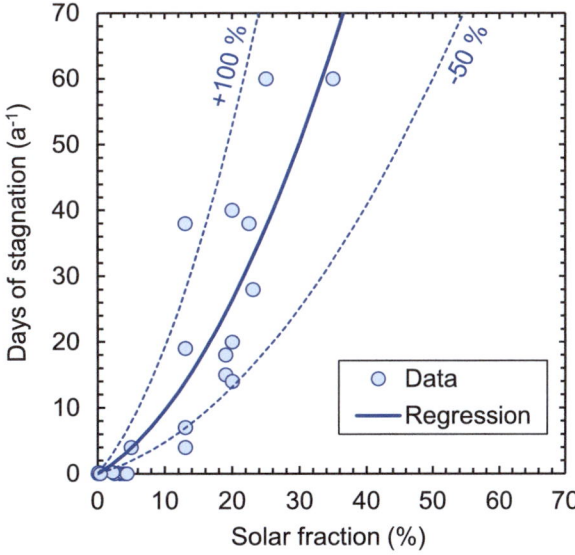

Fig. 3 Number of annual stagnation days in relation to the solar fraction based on 24 data sets

4 Results from Monitoring

The considered SDH features a collector field with approx. 6000 m^2 of ETC supplying a district heating network with an annual heat demand of about 16 GWh/a located in the middle of Germany. The targeted network supply temperature is between 75 and 90 °C depending on seasonality. The total TES capacity is only 40 l/m^2, which is significantly below the average of all the analyzed systems (see Fig. 2). The measurement setup, e.g. using meteorological sensors and heat energy meters, enables a fully energetic evaluation. Besides the collector field other heat generators, such as gas boiler and CHP units (combined heat and power), also load the multifunctional TES and the heating network (see Fig. 4). The control of the solar circuit (primary and secondary) is largely based on irradiation and primary supply temperature. The load management between SDH part and GHP units corresponds more to a manual system control by the operator, which results in a displacement of each other (depending on weather conditions). The measured annual yield was determined to 373 kWh/m^2 and represents the actually fed-in energy (including heat losses due to distance between collector field and feed-in point). Thus, the SDH system was able to supply 13% of the total heat demand (SF). The monitoring was carried out for the first year after commissioning, so there were also some data recording failures. Therefore, the focus of our analysis is especially on the evaluation of stagnation events.

Already during the planning phase of this SDH system, a relevant number of stagnation events was predicted due to the comparatively low TES capacities. Stagnation is expected to occur particularly during periods of low heat demand, especially in the summer months. Additionally, suboptimal control strategies can promote stagnation. An exemplary stagnation day is illustrated in Fig. 5, showing temperatures, irradiance, and thermal loads. During the day, the temperatures of the solar circuit (supply

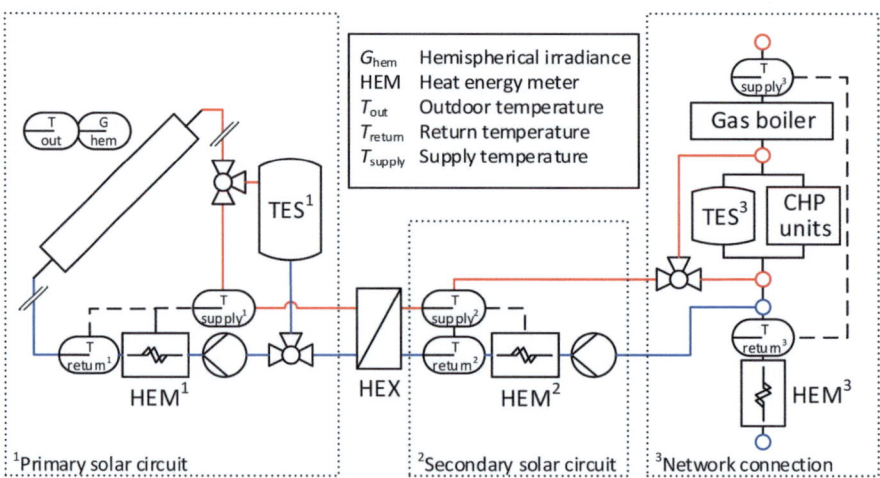

Fig. 4 Simplified measurement and hydraulic scheme

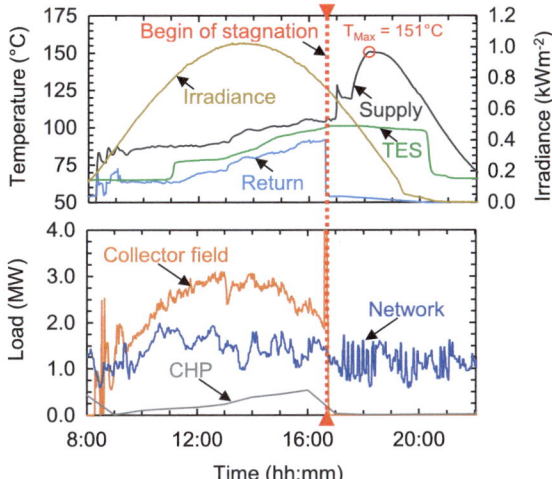

Fig. 5 Collector temperatures (supply and return), TES temperature and solar irradiance as well as load of solar thermal, CHP and network demand for an exemplary stagnation day

and return, measured at the pump station) as well as the TES temperatures are rising because the thermal power is significantly higher than the network's demand. At about 4:45 pm, the maximum temperature in the TES is reached, resulting in shutdown of all heat sources. Therefore, stagnation occurs in the solar thermal system and leads to draining of the collectors with temperature loads of 151 °C.

For the considered year, we identified a total number of 38 stagnation days. Due to some data recording failures, the actual number is probably even higher. Based on our results, the operator has extended the TES capacities by 35 l/m^2 to a total volume of 75 l/m^2. Further optimization can be achieved by an improved control strategy. The CHP's often feed-in, although there is already a significant surplus from the collector field (see Fig. 5). A more automated control strategy with well-defined priorities was clearly recommended and currently in implementation now. Furthermore, a predictive controller considering the weather forecast can also optimize the system operation.

5 Conclusion

This study investigates the integration of solar thermal energy in district heating systems (SDH) in Germany. Based on an analysis of feasibility studies and realized projects, the following conclusions can be drawn:

- Supply temperatures of 75–80 °C are common in Germany's SDH systems and can be effectively supplied by solar thermal collectors.
- Solar thermal, in combination with other renewable sources such as biomass, plays a crucial role in decarbonizing these networks, despite occasionally high temperature requirements.

- Stagnation becomes increasingly relevant in large-scale systems, particularly when thermal storage capacity is limited. Solar fractions of around 20% are associated with up to 50 stagnation days per year, which requires effective mitigation through larger storage volumes, advanced control strategies, or other stagnation prevention measures.

The results represent the current status of large-scale SDH networks in Germany. The observed quantities of stagnation events underline the fact that an intrinsically safe and holistic approach for stagnation handling is necessary for large-scale SDH systems. In addition to risk reduction, there are also potentials for cost savings by using less temperature stable but also less expensive materials in the solar circuit, e.g. polymeric pipes. Further savings are to be expected in the planning and implementation stages. In the district heating sector, the saving potential of polymeric piping systems is estimated with around 30% of the total piping costs, depending on individual conditions (e.g. ground properties and pipe diameter) [8]. In the R&D project *HP-BIG*, we are developing heat pipe collectors with inherent temperature limitation as large-area modules in an innovative system concept. The implementation of a demonstration plant as well as the comparison of the resulting costs for components and installation are the focus of the current work.

Acknowledgements The project "Reduction of the heat price of large-scale solar thermal systems with heat pipe vacuum tube collectors" (reference number 03EN6011A-C) underlying this publication, was funded by the state of Lower Saxony and the German Federal Ministry of Economy and Climate Action, following a decision of the German Parliament. The authors are grateful for this funding. The investigations were carried out in cooperation with the companies *AKOTEC Produktionsgesellschaft mbH*, *NARVA Lichtquellen GmbH + Co. KG* and *empact GmbH*. The authors are grateful for this support. The responsibility for the content of this publication lies with the authors.

References

1. Abrecht S et al (2024) Solar collector technologies for district heating. Report from IEA SHC Task 68. https://doi.org/10.18777/ieashc-task68-2024-0002
2. Hay S (2022) Existing district heating networks in context of german climate goals: potentials for UrbanTurn. In: ISEC conference Graz
3. Tschopp D et al (2020) Large-scale solar thermal systems in leading countries: a review and comparative study of Denmark, China, Germany and Austria. Appl Energy. https://doi.org/10.1016/j.apenergy.2020.114997
4. Spörk-Dü M et al (2025) Solar heat world wide. In: global market development and trends 2024. Detailed Market Figures 2023
5. Schiebler B et al (2023) Heat pipe collectors with overheating prevention in a cost-optimized system concept: monitoring of system performance and stagnation loads under real conditions. Solar Energy Adv 3:100040. https://doi.org/10.1016/j.seja.2023.100040
6. Fink C et al (2025) Wissenschaftliche begleitforschung zum förderprogramm Accessed at Mai 09 2025"solarthermie—Solare Großanlagen" 2010–2015. https://solarthermalworld.org/wp-content/uploads/2017/03/solare_grossanlagen_2010-2015_aggregierter_bericht.pdf. Accessed at 20 Mai 2025

7. Meißner R (2022) Technische herausforderungen mit grossflächen-solarthermie für die fernwärme. In: Conference proceedings of symposium solarthermie und innovative wärmesysteme. Bad Staffelstein, Germany, pp 273–283
8. Rehau: Effizienz-Technologie Nahwärme. Die Zukunft von Raumheizung und Warmwasser, https://www.rehau.com/downloads/404296/kommunale-waermeinfrastruktur-referenzen.pdf. Accessed at 09 Mai 2025

Open Access This chapter is licensed under the terms of the Creative Commons Attribution 4.0 International License (http://creativecommons.org/licenses/by/4.0/), which permits use, sharing, adaptation, distribution and reproduction in any medium or format, as long as you give appropriate credit to the original author(s) and the source, provide a link to the Creative Commons license and indicate if changes were made.

The images or other third party material in this chapter are included in the chapter's Creative Commons license, unless indicated otherwise in a credit line to the material. If material is not included in the chapter's Creative Commons license and your intended use is not permitted by statutory regulation or exceeds the permitted use, you will need to obtain permission directly from the copyright holder.

Enabling Data Economy for DHC Applications Through Dataspaces

Edmund Widl, Christian Wolff, Mathieu Vallée, Kristina Lygnerud,
Dietrich Schmidt, Zheng Grace Ma, Xiaochen Yang, Axel Oliva,
and Michele Tunzi

Abstract Digitalization enables innovative applications in district heating and cooling (DHC), supporting decarbonization and sector integration. These applications rely heavily on data, making the shift toward a data economy a logical next step. A data economy would let stakeholders collect, share, and monetize data, encouraging sustainable innovation. However, limited data availability, interoperability challenges, and unclear business models present key barriers. A potential solution may be dataspaces, which promote collaboration, innovation, and regulatory compliance while protecting sensitive data. This study presents DHC use cases showing how dataspaces can enhance operational flexibility, improve interoperability, and unlock new business value, offering a path to overcome digitalization challenges in the DHC sector.

Keywords Data economy · Dataspaces · Data access · Interoperability

---

E. Widl (✉)
AIT Austrian Institute of Technology, Vienna, Austria
e-mail: Edmund.Widl@ait.ac.at

C. Wolff · A. Oliva
Fraunhofer ISE, Freiburg, Germany

M. Vallée
Univ. Grenoble Alpes, CEA, Liten, Campus Ines, Le Bourget du Lac, France

K. Lygnerud
Lund University, Lund, Sweden

D. Schmidt
Fraunhofer IEE, Kassel, Germany

Z. G. Ma
University of Southern Denmark, Odense M, Denmark

X. Yang
Tianjin University, Tianjin, China

M. Tunzi
Technical University of Denmark, Kongens Lyngby, Denmark

© The Author(s) 2026
D. Vanhoudt (ed.), *Proceedings of the 19th International Symposium on District Heating and Cooling*, Lecture Notes in Networks and Systems 1700,
https://doi.org/10.1007/978-3-032-09844-3_5

1 Introduction

Digitalization facilitates a wide range of innovative applications for district heating and cooling (DHC), which are expected to become important enablers for the heat sector's decarbonization and integration with other sectors [11]. The aim and scope of these new applications vary, but they all require access to data. Given this dependence on data, the transition towards a data economy in DHC is the logical next step for the successful implementation of these applications. However, limited data availability (e.g., due to regulations on information privacy) and data interoperability (e.g., due to proprietary data formats) as well as knowledge on what business model implications the data economy has are often critical barriers.

The European Commission announced in the European Strategy for Data its intent to create a single market for data sharing and exchange that is efficient and secure. Dataspaces were identified as the central key technology, facilitating data exchange by enabling collaboration and innovation, while also protecting sensitive data and simplifying regulatory compliance [2]. Dataspaces are expected to become ubiquitous within the European Union (like clouds are today), enabling data access across all sectors, including transport, finance, health, and governance. The *Common European Energy Data Space* [1] is supposed to establish common markets for energy and data to help facilitate the energy transition.

Assuming that dataspaces will become a common digital infrastructure, they might also be utilized for breaking down data silos that impede the implementation of a data economy in the DHC sector and improve interoperability of products from different vendors. Within this context, this study presents use cases that demonstrate the advantages of using dataspaces for DHC, addressing both operational improvements through flexibility and other, added business values.

2 Opportunities of Data Economy in DHC

The term *data economy* refers to a (global) digital ecosystem where data is gathered, organized and exchanged to create economic value. Within the context of DHC, moving towards a data economy has the potential to create value that can be capitalized on in several areas. This includes the improvement of operational efficiency, the reduction of fossil fuel use, the increase of flexibility and resilience of the system, customer engagement, as well as new business models expanding into new regions or customer segments. Hence, active participation in a data economy can have specific benefits for many stakeholders, described in the following.

DHC network operators can create value by leveraging data to improve the operational efficiency of the system. Utilization of real-time and historical data enables advanced operational optimization to significantly cut down on downtime and energy losses, e.g., via predictive maintenance, load forecasting, and leakage detection. The sharing of data with other stakeholders (e.g., power network operators) enables improved coordination, particularly in integrated energy systems. The transformation towards renewable and fluctuating energy sources will make digitalization and data integration a mandatory process to operate DHC networks in future. Furthermore, data exchange with other actors is a requirement for cross-sectorial planning of infrastructure.

Energy service companies can use data to create revenue by offering customized services and performance optimizations to DHC customers. For instance, analysis of customer energy usage data can be used to design customized savings and plan retrofit projects both on the grid side and on the customer side (e.g., for reducing return flow temperatures). At the same time, the data can be used for remote monitoring and diagnostics of performance to reduce on-site visits and improve service quality, e.g., via automated data-driven adjustments of controller parameter to optimize consumption patterns and costs.

Technology/software providers can integrate operational and customer data into their R&D process for improving their software tools for energy management, demand forecasting, and load balancing. They can also act as enablers for a data economy by building and operating interoperable platforms that aggregate and analyze data from various stakeholders (smart meters, smart appliances, IoT sensors, etc.) and provide advanced analytics tools (e.g., AI and machine learning services) on top of it that help operators or energy service companies make smarter decisions.

3 Challenges for Data Access and Interoperability in DHC

Barriers to data sharing in the DHC domain can be categorized into *technical*, *organizational*, and *regulatory* challenges. Although technical barriers are often the first to be mentioned, experience suggests that organizational and regulatory barriers are just as important in the DHC sector.

Organizational Barriers

- *Competitive advantage concerns*: Sharing data is often perceived as a loss of competitive advantage. Particularly when networks are operated under concession agreements (and operators do not own them) sharing data about a network, its operation, and its clients, could help competitors better position themselves during the renewal of concessions [6]. Similarly, sharing data with the municipality or other network owner provides them with a means of auditing, which some companies refuse in the absence of obligation.
- *Loss of control*: Data is often perceived as a potential revenue generator, so that sharing can be perceived as a loss of control and create a fear of missing opportunities [4]. Even within organizations that could benefit from data sharing to improve their operations, different teams may resent to do so, especially when missing a clear structure for data sourcing and responsibility.

- *Resource and competence constraints*: The lack of resources or skills to manage data represents a significant barrier. Collecting, cleaning, and providing data requires time and qualified resources [12]. In the absence of a clear model for return on investment (ROI) associated with this sharing, companies hesitate to make these efforts.

Regulatory Barriers

- *Compliance*: Companies may fear financial sanctions and reputational damage in case of non-compliance or mismanagement of shared data. For example, legal risks related to personal data protection (GDPR–General Data Protection Regulation) are a significant concern [5]. The lack of internal expertise in data management can amplify this phenomenon.
- *Data ownership ambiguity*: In some cases, data ownership is loosely defined, especially when data are derived from other data provided by external actors [7]. The lack of clear regulations on the rights and duties regarding data sharing creates a legally uncertain environment.

Technical Barriers

- *System Heterogeneity and vendor lock-in*: Energy management solution providers often use proprietary data formats, making integration and information exchange between systems difficult. This technological disparity requires additional efforts to harmonize protocols and standardize data formats, which may require considerable costs and time [12].
- *Data quality*: Collected data can be incomplete, inaccurate, or non-standardized, making the sharing and use inefficient. For example, relevant data quality standards [10] are required to avoid interpretation errors or even impossibility to interpret the data.
- *Data security*: Concerns about data security are a major barrier to information sharing [8]. The risks of cyberattacks or leaks of sensitive information can deter actors from sharing their data, even when it would be beneficial.

Efforts to facilitate data access and interoperability have been more limited in the DHC sector compared to the power system sector. In this context, the introduction of the energy dataspace approach is particularly promising for DHC.

4 Introduction to Dataspaces

Dataspace is a collective term used to describe emerging concepts and technologies that aim at addressing the challenges of data availability, data interoperability, and data semantics. Standardization initiatives are currently ongoing, however, this process is not yet finished, resulting in a multitude of definitions.

In general, however, there is agreement that a dataspace refers to a structured environment for storing, managing, and accessing data across multiple sources and systems within an organizational, national, or international context. Dataspaces encompass the frameworks, policies, standards, and technologies that enable the secure, efficient, and interoperable exchange and processing of data among diverse stakeholders. Dataspaces are designed to support a wide range of applications, from business intelligence and analytics to machine learning and artificial intelligence, by providing a coherent and integrated approach to data management [3].

Dataspaces can start relatively simple yet can evolve to be quite complex. Still, it is reasonable to expect at least some basic functionalities to exist:

- *Identity management*: covers the registration of participants, assigning them roles that they are capable and willing to perform within the dataspace
- *Infrastructure functionalities*: capabilities that enable the secure and trusted exchange and sharing of data in a manner that respects privacy, security, and intellectual property rights
- *Intermediaries*: capabilities that enable dataspace participants to connect to the infrastructure
- *Vocabulary provider*: capabilities to manage vocabulary specifications for the semantic interpretation of data provided through the dataspace

The *Data Spaces Blueprint* [2] provides a detailed view of the typical building blocks of a dataspace. Notably, these building blocks are not limited to technical concepts and components, but also encompass organizational elements targeting business, governance and legal aspects. According to the *Reference Architecture Model for the International Data Spaces* (IDS-RAM) [9], the total number of its so-called *connectors* is what constitutes a dataspace. Connectors are dedicated communication servers for sending and receiving data in compliance with the IDS-RAM's specification. The data endpoints that are exposed by each connector make it possible for different data to be exchanged. As a consequence, dataspaces enable the federated storage of data, without the need for a central data storage.

5 Use Cases for Data Economy in DHC Enabled by Dataspaces

In the following, use cases are presented that showcase how dataspaces could help address the opportunities and challenges discussed above. These use cases by no means represent an exhaustive list, but rather serve as a starting point for exploring the potential of dataspaces for the DHC domain. The advantages of dataspaces stated for each use case are based on experience from other domains. Future R&D is required to obtain empirical proof supporting these statements.

5.1 Use Case 1: Improving Operational Efficiency with Real-Time Customer Data

Background Network operators often lack access to real-time or predictive data from customer premises, such as building-level heat demand, indoor temperatures, or thermal storage capacities. Instead, operators have to rely on aggregate demand estimates or outdated consumption patterns. This lack of granular and timely information limits their ability to implement demand-side management and flexible operation strategies. This causes inefficiencies such as overproduction, heat losses, and suboptimal distribution, especially during demand peaks or variable weather conditions. However, due to privacy concerns, regulatory constraints, decentralized ownership of data, and a lack of machine-to-machine interfaces, operators often cannot access the desired information directly.

Scenario A dataspace provides a federated, secure mechanism to access and share customer data without breaching ownership or privacy, enabling data-driven operational optimization. Smart buildings in the network may share data (e.g., real-time heat demand forecasts, thermal storage availability, and indoor temperature flexibility ranges) via the dataspace through smart contracts. Based on this information, the operator can reduce heat losses by minimizing over-distribution and integrate variable renewable sources more effectively. In addition, the operator may send flexibility requests to certain customer segments via the dataspace. Based on data-driven algorithms, the operator can identify buildings with thermal inertia that can pre-heat during off-peak hours and dynamically adjusts supply temperatures in the network. This coordinated system allows for real-time demand shaping, peak shaving, and optimal use of variable heat sources, all without direct control over customer premises.

Advantages Offered by Dataspaces

- *Data sovereignty and trust*: Customers retain control over their data, deciding who can access it, under what conditions, and for which purposes.
- *Interoperability and standardization*: Standardized data formats and interfaces allow to streamline the integration of various metering, control, and forecasting systems.
- *Granular access control*: Only the required data is accessed (e.g., hourly consumption patterns, aggregated by zone) without accessing personally identifiable information, ensuring compliance with privacy laws.
- *Scalability*: As more buildings and services join the dataspace, the system becomes more responsive and efficient without the need for centralized data storage or processing.

Barriers Digital ecosystems around DHC systems are typically quite complex, covering many areas (operation, billing, etc.). Upgrading all the relevant systems is a complex and costly task, involving many stakeholders and software vendors.

5.2 Use Case 2: DHC Software Market Liberalization

Background A range of software tools has become available to help manage decentralized heat sources, grid transformation, and expansion. These can be utilized for a variety of purposes, including operational optimization, leakage detection, expansion planning, and monitoring. However, all these tools need access to data from the DHC systems, requiring the availability of (machine-to-machine) interfaces to set up the software. The development of these interfaces can be expensive and may diminishing economic efficiency. Furthermore, due to the lack of standardized interfaces and data formats, the development of these bespoke interfaces can also create a dependency of the network operator on the software vendor.

Scenario The standardization of dataspaces through initiatives like the IDS-RAM will provide uniform access to a wide spectrum of data sources. For network operators, this eases the provision of access to relevant data to external stakeholders. Hence, network operators can request dataspace compatibility in tenders for software providers. This saves software providers time and effort, as they do not have to develop customized interfaces and can price their services more effectively. It also makes it easier for network operators to choose between several software providers for different services. As a result, full-range providers lose the advantage of only having to write one interface for all applications and the market becomes more interesting for smaller specialized applications. It also reduces the risk of vendor lock-in, as it is possible to switch to other software providers at lower cost.

Advantages Offered by Dataspaces

- *Interoperability and standardization*: Dataspaces will provide uniform access to a broad spectrum of data sources, preventing vendor lock-in due to standardized interfaces.
- *Market liberalization*: Specialized software can be easily connected to existing data.
- *Economic efficiency*: Software tools can be implemented with less effort, enabling cheaper solutions.

Barriers The barriers to market liberalization concern the prevailing data models within organizations and the considerable effort required to develop new interfaces between the dataspace and software providers. Furthermore, partnerships between software providers and network operators are often long-term, relying on already established interfaces.

5.3 Use Case 3: Third-Party Access to Heat Supply

Background The transformation to renewable energy sources is a difficult endeavor for DHC network operators. This often requires large investments in new heat plants and the grid infrastructure. Third-party access to renewable heat sources and unavoidable waste heat would enable the mobilization of private capital, alleviating the strain on public budgets while also accelerating decarbonization. However, apart from regulatory aspects, operational issues must also be addressed, as the exchange of data and the transparency of heat supply costs plays a key role here.

Scenario A network operator has several co-operative, industrial and private heat supply systems in its network. Each must be integrated differently according to their contract. The industrial plants provide a certain amount of waste heat that must be purchased, whereas co-operative plants compete with the operator's own plants. Hence, there must be a data exchange to ensure that each plant can be operated optimally and at the same time can be regulated via the plant's price signal. Dataspaces enable the easy and secure exchange of data and the rapid integration of different plants into the optimization problem, both in terms of technical aspects (e.g., interfaces) as well as governance structures (e.g., smart contracts).

Advantages Offered by Dataspaces

- *Interoperability and standardization*: Dataspaces enable the communication between different plants, using common interfaces to bridge the gap between different technologies.
- *Market liberalization*: Standardized interfaces ease the access of third party heat providers through common governance structures.
- *Scalability*: New systems and the conversion of existing systems could be realized more quickly thanks to the new capital from private investors in heating plants.

Barriers The biggest challenge in allowing third-party access to heat supply systems is the lack of regulations and advanced control algorithms.

6 Conclusions

This work explores how dataspaces can facilitate the development of a data economy in the DHC sector, addressing key barriers such as limited data availability, interoperability challenges, and unclear business models. By enabling secure, standardized, and privacy-preserving data exchange, dataspaces allow diverse stakeholders to access and leverage real-time and historical data. This supports improved operational efficiency, system flexibility, and the integration of renewable energy sources. Use cases are presented highlighting their potential to overcome digitalization challenges and unlock new value in the DHC sector.

References

1. Berkhout V, Villeviere C, Bergsträßer J, Klobasa M (2023) Common European energy data space. Publications Office of the European Union. https://doi.org/10.2833/354447
2. Data Spaces Support Center: Data Spaces Blueprint v2.0. Tech. rep. https://dssc.eu/space/BVE2/1071251457/
3. European Commission (2020) A European strategy for data. https://eur-lex.europa.eu/legal-content/EN/TXT/?uri=celex:52020DC0066
4. Jones CI, Tonetti C (2020) Nonrivalry and the economics of data. Am Econ Rev 110(9):2819–2858. https://doi.org/10.1257/aer.20191330
5. Lee D, Hess DJ (2021) Data privacy and residential smart meters: comparative analysis and harmonization potential. Util Policy 70:101188. https://doi.org/10.1016/j.jup.2021.101188
6. Li F et al (2025) Do we need a data sharing infrastructure for the energy sector? IET Smart Grid 8(1):e12196. https://doi.org/10.1049/stg2.12196
7. Lina J, Ting W, Xiaohua W (2022) Research and design on the confirmation method of power data assets. In: Big Data—BigData 2022. Springer International Publishing, pp 61–69. https://doi.org/10.1007/978-3-031-23501-6_7
8. Mollah MB et al (2021) Blockchain for future smart grid: a comprehensive survey. IEEE Internet Things J 8(1):18–43. https://doi.org/10.1109/JIOT.2020.2993601
9. Otto B et al (2025) IDS reference architecture model. https://docs.internationaldataspaces.org/ids-knowledgebase/ids-ram-4
10. Schaffer M, Tvedebrink T, Marszal-Pomianowska A (2022) Three years of hourly data from 3021 smart heat meters installed in Danish residential buildings. Sci Data 9(1):420. https://doi.org/10.1038/s41597-022-01502-3
11. Schmidt D (ed) (2023) Guidebook for the digitalisation of district heating: transforming heat networks for a sustainable future: final report. AGFW-Project Company. ISBN 978-3-89999-096-6
12. Wang J et al (2023) Data sharing in energy systems. Adv Appl Energy 10:100132. https://doi.org/10.1016/j.adapen.2023.100132

Open Access This chapter is licensed under the terms of the Creative Commons Attribution 4.0 International License (http://creativecommons.org/licenses/by/4.0/), which permits use, sharing, adaptation, distribution and reproduction in any medium or format, as long as you give appropriate credit to the original author(s) and the source, provide a link to the Creative Commons license and indicate if changes were made.

The images or other third party material in this chapter are included in the chapter's Creative Commons license, unless indicated otherwise in a credit line to the material. If material is not included in the chapter's Creative Commons license and your intended use is not permitted by statutory regulation or exceeds the permitted use, you will need to obtain permission directly from the copyright holder.

Mitigating the Grid Impact of Electrifying Heat: A Case Study Using Solar and Thermal Storage to Achieve a Net-Zero Emission Campus

Eric Ho, Raymond Boulter, and Jeff Thornton

Abstract This study investigates the technical feasibility of achieving carbon neutrality in campus heating through integration of on-site PV or solar thermal and thermal storage systems. Using a simulation case study based on a real, federally owned campus in Ontario, Canada, various scenarios were explored to assess their potential to meet Canadian 2050 net-zero emission targets while minimizing electrical demand. Results indicate that multiple scenarios employing a combination of PV, solar thermal, BTES, and PTES on a district energy system can achieve carbon neutrality. However, variations in annual consumption, peak demand, and land use highlight the trade-offs that exist in evaluating the feasibility of net-zero emission campuses.

Keywords Campus · Solar · Thermal storage · Land use · Electrical demand

1 Introduction

1.1 Background

To mitigate the impending 1.5 °C temperature rise, the United Nations (UN) has set a global target of achieving net-zero emissions by 2050 [1], and many countries, including Canada, have set similar national targets. The Canadian government also updated their Greening Government Strategy (GGS) in Fall 2020, establishing new directives and reduction targets specifically for Canadian federal operations [2]. This includes reducing GHG emissions from federal facilities and fleets by 40% by 2025 and at least 90% by 2050. To help meet these ambitious targets, federal campus

E. Ho · R. Boulter (✉)
Natural Resources Canada, CanmetENERGY-Ottawa, Ottawa, Canada
e-mail: raymond.boulter@nrcan-rncan.gc.ca

J. Thornton
Thermal Energy System Specialists (TESS), Madison, USA

and building operators can potentially replace their fossil fuel heating systems with those that use electricity only, since most provincial electricity grids in Canada are projected to have near zero emissions by 2030 [3]. However, the grid projections are based on annual energy consumption and do not consider the full time-of-use implications of electrifying heat loads [4]. Consequently, there may be insufficient zero-emission generation, transmission, and distribution capacity available to meet the required heating demands.

1.2 Previous Studies and Research Gaps

Grid interactions and low-emissions energy systems for stand-alone buildings have been explored extensively. These systems often comprise renewable on-site generation and some form of energy storage controlled by an energy management system (EMS). A large focus of research is on the EMS models, in which the objective is to ensure that annual energy consumption is met by the generation on-site, ultimately achieving "net-zero energy" [5]. With the advent of net-zero energy building (NZEB) design, evaluations beyond the annual energy balance have been increasingly considered, such as the impact on the electrical grid, GHG emissions, and land use [6].

While NZEB design primarily revolves around demand response strategies to align load with on-site generation [7], a more comprehensive approach is required to address the unique nature of campuses. Campuses are made up of multiple interconnected buildings; they must consider the collective impact of their buildings on the grid and address peak electrical demand issues. Moreover, campuses have high incentives to reduce their electrical demands: they must work with utilities to ensure there is sufficient capacity in distribution and substations to support the aggregated demand of all buildings, and must pay charges for electrical demand in addition to consumption [8]. Recent investigations at the urban and neighbourhood level have demonstrated the potential for on-site renewables and thermal storage to reduce GHG emissions in campuses with district energy (DE) systems [9–11]. However, limited studies—primarily focused in Europe—have addressed the crucial issue of peak demand [12, 13].

1.3 Objective

Previous carbon neutrality studies with renewable energy and storage have either included minimal investigation of grid implications or have not been applied at the campus scale. Using a simulation-based study, this paper aims to:

Table 1 Case study campus building summary

Building type	Total floor area (m^2)	Number of buildings
Maintenance shop	10,213	1
Office	63,148	2
Storage	7442	2
Training	1971	1
Vehicle garage	19,140	4
Total	101,914	10

- Demonstrate if photovoltaic (PV) and solar thermal technologies with thermal storage on a campus DE system in Ontario, Canada could achieve net-zero emissions while minimizing electrical demand.
- Explore the advantages to a DE system of different configurations of PV and solar thermal technologies with thermal storage.
- Investigate metrics that are useful to evaluate the feasibility of a net-zero emission campus, including electrical peak demand and land use.

2 Methodology

2.1 Case Study

As there are no examples of large-scale solar and thermal energy storage systems in operation on campuses in Canada, a simulation-based case study was effectuated for the purpose of this analysis. The case study uses a federal government campus located in Ontario, Canada. At present, the campus does not have a district heating plant, solar or thermal storage. Some site specifications have been anonymized, and no building layout map has been included here per owner request. A list of the types and floor areas of the campus' buildings is given in Table 1. All were built between 2014 and 2017. A variety of scenarios were devised to align with the overarching goal of achieving net-zero emissions.

2.2 Energy Load Modelling

Due to the lack of metered data available at the building level, an archetypal building energy model approach was used to model the total energy load of the campus without conducting full audits of the buildings. The building models used were chosen from an EnergyPlus[1] directory of federal building archetypes developed by

---

[1] EnergyPlus™ whole building simulation software, available at https://energyplus.net/.

Natural Resources Canada using real buildings as references. The directory is closed-sourced due to security considerations and special permission was obtained for this study. Archetypes were assigned according to "Building type" in Table 1 and the 2005–2014 vintage. Typical Meteorological Year weather data from the Canadian Weather Year for Energy Calculation (CWEC) were used to simulate each archetype using EnergyPlus on an hourly interval. From each archetype, the hourly space and water heating loads and plug electric loads were extracted and normalized by floor area to obtain energy use intensity (EUI). The per area loads were multiplied by the area of each building on the campus to estimate the building energy loads. Following iterative adjustments related to archetype assignments, along with considerations for factors such as non-building energy loads (e.g., streetlights, exterior lighting), the modelled loads were validated to within +/−10% of the electric and gas utility data for the entire campus.

2.3 District Energy System Modelling

First, to model the DE system, a physical layout of the district piping, strategically laid according to building locations, was created. To optimize efficiency, a central heating plant was positioned in an open space in proximity to the heating network. The size of network pipes was determined using the RETScreen[2] hot water piping sizing tool, factoring the space and water heating loads obtained from EnergyPlus simulations.

Subsequently, the operations and performance of the DE system and technology scenarios were modelled in TRNSYS, with the aid of the TESS Component Library Package.[3] This approach captures the comprehensive dynamics of the DE system, accounting for variables such as heat transfer, heat loss, and overall system efficiency. The thermal network was designed at 70 °C with a temperature difference of 20 °C. The network operates with an outdoor reset scheme with a supply temperature of 70 °C for ambient temperatures − 10 °C and lower, at 50 °C for ambient temperatures above 10 °C, and with a linear supply temperature relationship factored for ambient temperatures in between. Although the TESS Component Library Package and TRNSYS are one of the industry standards for system simulations, due to the fact there is no current DE system on the site, there is no real data to validate the DE model. While this is a limitation, the models are only analyzed comparatively under the same assumptions.

---

[2] RETScreen® Clean Energy Management Software platform available at https://natural-resources.canada.ca/maps-tools-publications/tools-applications/retscreen.

[3] TRNSYS and TESS Component Library Package available at https://www.trnsys.com/.

2.4 Scenario Descriptions

At the current campus, all buildings are heated by decentral, building-level, natural gas boilers. This was modelled as the business as usual (BAU) scenario. Scenarios primarily using decentral electric boilers and decentral air source heat pumps[4] (ASHP) were modelled as M1 and M2, respectively. Centralized district heating scenarios (M3 to M10) were developed to allow the use of large-scale solar thermal collectors (STC) and PV, along with water source heat pumps (WSHP), and thermal energy storage (TES), including short term storage (STTES), and seasonal borehole (BTES) and pit (PTES) storage. General parameters considered for the key technologies are summarized below:

BTES (TRNSYS type 1373): boreholes were assumed to be 100 m deep, 0.15 m in diameter, have 2.25 m spacing, and use 1–1/4″ SDR-11 HDPE U-tube piping. Surface insulation of R-40 (ft2·°F·h/Btu) extending 25 m beyond the storage volume was assumed. BTES fields were sized based on an assumed relationship of 2 m of total length per m^2 of STC or PV panel. An initial earth temperature of 5.74 °C was assumed, with soil thermal conductivity of 7.50 kJ/h-m–K and capacitance of 2000 kJ/m^3-K.

PTES (TRNSYS types 1535 & 1301): modelled as a conical tank with soil wrapper having a center depth of 16 m and a ratio of top radius to bottom radius of 2.15. An R-40 insulated cover and R-40 surface insulation extending 25 m around the PTES perimeter was assumed. The PTES volume was sized using a relationship of 100 m^3 per STC and 16 m^3 per PV panel.

STTES (TRNSYS type 1534): included where needed to act as a buffer for charging and discharging thermal energy to and from the BTES, modelled as a 3000 m^3 vertical insulated tank.

Top-up storage tank (TRNSYS type 1534): included where needed to help meet peak DE demands, modelled as a 500 m^3, vertical insulated tank, charged to 90 °C.

STC (TRNSYS type 1346): ground-mounted solar collectors used for generating hot water to charge the PTES or BTES (via STTES tank) or to directly meet the DE load. STCs were assumed to be double glazed, large format (13 m^2/module), generally based on the GREENoneTEC[5] GK3003 collector. Collectors faced due south at 45° slope.

Photovoltaics (TRNSYS type 103): ground-mounted solar panels used for producing electricity for grid export, to offset electrical energy used by the heat pumps and boilers. Each PV panel was assumed to be 440 W, 2.21 m^2, and face due south at 45° slope.

WSHP (TRNSYS type 1399): heat pumps were modelled as multiple modular CO_2 (R744) units in each scenario, and assumed to operate similarly to a Mayekawa

---

[4] Decentral ASHP COP assumed to range between 2.2 at − 16 °C and 3.7 at 16 °C., with electric resistance supplementary heating (COP 1.0) applied for ambient temperatures below − 16 °C.

[5] GREENoneTEC: https://www.greenonetec.com/en/units/large-scale-projects/products/

Unimo[6] heat pump, to generate hot water up to 90 °C for the DE network and for charging the PTES or BTES and STTES tanks.

Electric boiler (TRNSYS type 659): used as needed to maintain the required DE supply temperature.

A summary of the key components included in the design of each scenario is pro-vided in Table 2. Note that the component capacities in the renewable system scenarios (M4 to M10) were sized through an iterative modelling process to achieve a peak electrical demand reduction of approximately 50% compared to scenario M3. Scenario equipment capacities should not be interpreted as being optimized.

3 Results

The modelling completed for each technology scenario produced outputs of the projected energy consumption by fuel type and associated GHG emissions. Figure 1 provides a comparative energy and GHG emissions summary of all scenarios in 2050. Given the use of (carbon-free) electric and solar energy sources, all scenarios achieve carbon neutrality. With the exception of Scenario M10, all solar scenarios use less energy than their all-electric scenario counterparts (M1, M2, and M3).

Figure 2 gives the projected monthly peak electrical demand for all scenarios, based on the hourly modelling completed. Each coloured bar in the graphs represents the modelled peak electric demand for the entire base in each of the 12 months, to provide an indication of the expected change in peak demand throughout the year. The solar scenarios are able to meet the heating needs of the campus while reducing peak electric demand by approximately 5000–6500 kW compared to the all-electric scenarios, constituting an improvement of 40–50%.

Detailed cost analysis is not included within the scope of this paper. However, Fig. 3 includes notional utility cost impacts for each scenario based upon current electricity rates for energy (CAD $127.20/MWh) and demand (CAD $17.20/kW), as well as natural gas price of CAD $5.75/GJ and a shadow carbon price of $300/tonne (per the GGS directives) applied for the BAU scenario. As shown, with the exception of M10, all the renewable scenarios are expected to have lower annual utility costs compared to BAU, as well as the all-electric scenarios (M1, M2, M3). The results also highlight the importance of considering not only the electrical energy costs, but the electrical demand cost impacts as well.

To evaluate expected land use,[7] Fig. 4 illustrates the total area used by each technology in the scenarios alongside their annual electrical consumption and annual electrical billed demand (i.e., the sum of the monthly peak electrical demand). The data reveals a trend wherein lower annual electrical consumption correlates with

---

[6] Mayekawa Unimo: https://mayekawa.com/products/heat_pumps/

[7] For solar technologies, required land area was approximated as total STC or PV panel area. For storage, *PTES $(m^2) = 0.112 \times V_{PTES} - 7.495$* and *BTES $(m^2) = 5.05 \times$ #boreholes $+ 32.56$*. Other system components were assumed to have minimal incremental land use impact.

Table 2 Summary of modelled scenarios

	Decentral		Central							
	M1	M2	M3	M4	M5	M6	M7	M8	M9	M10
STC area (m^2)	–	–	–	79,000	36,200	–	52,700	32,900	–	23,000
PV area (m^2)	–	–	–	–	–	55,300	–	–	38,700	–
ASHP (kW$_t$)	–	14,300*	–	–	–	–	–	–	–	–
Electric boiler (kW$_t$)	11,747*	–	11,613	5717	–	5158	5237	327	5161	–
WSHP (kW$_t$)	–	–	–	–	10,840	14,551	–	11,329	12,827	10,603
BTES length (m)	–	–	–	39,500	18,100	27,650	–	–	–	–
PTES volume (m^3)	–	–	–	–	–	–	400,000	250,000	280,000	175,000
STTES volume (m^3)	–	–	–	3000	3000	3000	–	–	–	–
Top-up tank (m^3)	–	–	–	–	500	500	–	–	500	500

*Total Decentral Capacity

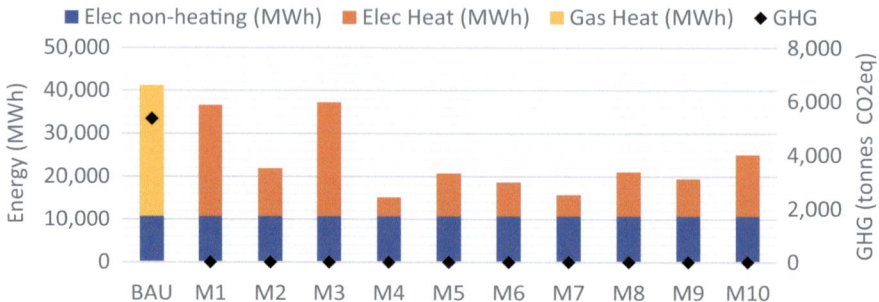

Fig. 1 Modelled scenario annual energy and GHG comparison

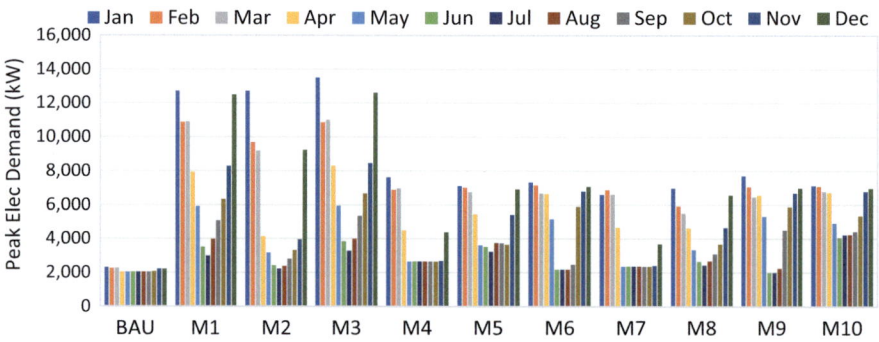

Fig. 2 Modelled total peak electricity demand by month for each scenario

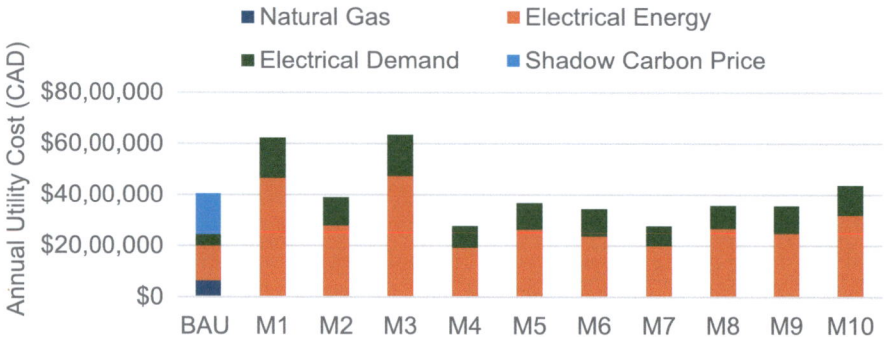

Fig. 3 Comparison of annual scenario utility costs, based on current rates

reduced billed demand, but at the cost of increased land use area. Notably, scenarios without the use of heat pumps (M4 and M7) exhibit the highest savings in both annual consumption and demand. However, these scenarios also have the highest land use, exceeding those that incorporate heat pumps by around one-third.

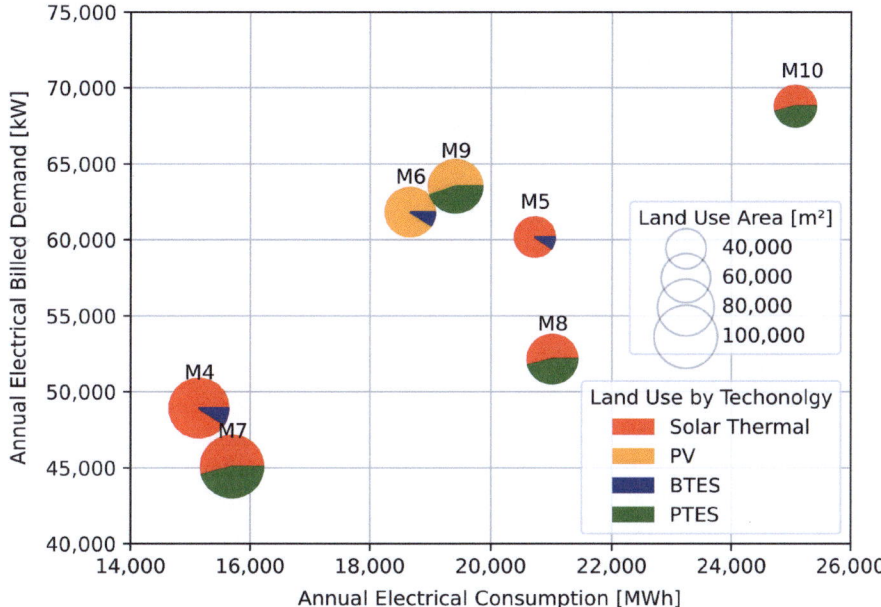

Fig. 4 Comparison of land use, electrical consumption, and billed demand (M4–M10)

4 Discussion

The modelling analysis confirms that solar energy systems along with thermal energy storage, when integrated within electric central heating plant configurations, can effectively mitigate overall peak electrical demand while achieving net-zero emissions (scenarios M4–M10). The analysis also revealed that scenarios utilizing BTES tended to require less land area for storage compared to those employing PTES, relative to the total land use for the scenarios. This difference likely arises from BTES systems' ability to utilize more vertical space underground. However, due to the higher heat capacity of water compared to soil, BTES scenarios require larger solar systems, nearly offsetting the land use savings from storage with the additional land required for the solar system. This trend is evident in the comparison between scenarios M6 and M9, as well as scenarios M4 and M7 in Fig. 4. Furthermore, scenarios M6 and M9 incorporate large-scale PV arrays, while the other renewable scenarios investigated incorporate large-scale solar thermal collector arrays. In the scenario design, it is assumed that the electricity produced from the PV panels would be used for heating. However, there is potential for the solar generated electricity to be used for any electric service on the campus, as a means of providing on-site generation for added energy resiliency.

The purpose of this study is not to definitively determine the best scenario, but to demonstrate that there are many pathways to achieve a net-zero campus. Since all scenarios deploying solar and storage were able to meet the net-zero emissions goal, additional metrics, including electrical peak demand and land use, were applied to carry out a more meaningful and comprehensive comparison. This exercise allows campus operators and owners to make informed planning decisions by prioritizing factors such as land restriction or high electrical demand costs. For example, if electrical consumption and demand are a priority, a solution with a large solar and thermal storage system without the aid of a heat pump can be used (similar to M4 or M7). On the other hand, if land use is the priority, systems like M5 or M10 where the heat pump can directly charge the storage using electricity would be favourable.

5 Conclusion

This study highlighted the many factors that should be considered to achieve net-zero emissions on campuses. By simulating different scenarios, the benefits of integrating solar energy and thermal storage into a district energy system were demonstrated. Notably, the analysis revealed the trade-offs between energy consumption, peak demand, and land use. While multiple decarbonization solutions were found to exist, the findings ultimately pointed to the importance of considering additional design factors before implementation. The next natural progression for this work would be to assess capital and operational costs through the completion of a life cycle costing analysis that includes consideration for capital expenditures, operation and maintenance, labour, and utility costs over an extended timeline (40 years is recommended within the Canadian GGS directive). As well, completion of sensitivity analysis on the simulated outputs would help to further validate the results.

References

1. IPCC (2018) Global warming of 1.5°C. Geneva, Switzerland
2. Treasury Board of Canada Secretariat (2017) Greening government strategy, Ottawa
3. Environment and Climate Change Canada Data Catalogue (2023) Electricity grid intensities. 07 Dec 2023. https://data-donnees.az.ec.gc.ca/data/substances/monitor/canada-s-greenhouse-gas-emissions-projections/Current-Projections-Actuelles/Energy-Energie/Reference%20Scenario%20de%20reference/Grid-O%26G-Intensities-Intensites-Reseau-Delectricite-P%26G/?lang=en
4. St-Jacques M, Bucking S, O'Brien W (2020) Spatially and temporally sensitive consumption-based emission factors from mixed-use electrical grids for building electrical use. Energy Build 224:110249. https://doi.org/10.1016/j.enbuild.2020.110249
5. Brahman F, Honarmand M, Jadid S (2015) Optimal electrical and thermal energy management of a residential energy hub, integrating demand response and energy storage system. Energy Build 90:65–75. https://doi.org/10.1016/j.enbuild.2014.12.039
6. Deng S, Wang RZ, Dai YJ (2014) How to evaluate performance of net zero energy building—a literature research. Energy 71:1–16. https://doi.org/10.1016/j.energy.2014.05.007
7. Tumminia G et al (2020) Grid interaction and environmental impact of a net zero energy building. Energy Convers Manag 203:112228. https://doi.org/10.1016/j.enconman.2019.112228

8. Brusco G, Burgio A, Menniti D, Pinnarelli A, Sorrentino N (2014) Energy management system for an energy district with demand response availability. IEEE Trans Smart Grid 5(5):2385–2393. https://doi.org/10.1109/TSG.2014.2318894
9. Guelpa E, Verda V (2019) Thermal energy storage in district heating and cooling systems: a review. Appl Energy 252:113474. https://doi.org/10.1016/j.apenergy.2019.113474
10. Wilk P, Cantor E, Li J (2023) Net-zero emission for multi-energy campus system. In: 2023 IEEE power & energy society general meeting (PESGM), pp 1–5. https://doi.org/10.1109/PESGM52003.2023.10252323
11. Del Borghi A, Spiegelhalter T, Moreschi L, Gallo M (2021) Carbon-neutral-campus building: design versus retrofitting of two university zero energy buildings in Europe and in the United States. Sustainability 13(16). https://doi.org/10.3390/su13169023
12. Holmér P, Ullmark J, Göransson L, Walter V, Johnsson F (2020) Impacts of thermal energy storage on the management of variable demand and production in electricity and district heating systems: a Swedish case study. Int J Sustain Energy 39(5):446–464. https://doi.org/10.1080/14786451.2020.1716757
13. Yan Z, Zhang Y, Liang R, Jin W (2020) An allocative method of hybrid electrical and thermal energy storage capacity for load shifting based on seasonal difference in district energy planning. Energy 207:118139. https://doi.org/10.1016/j.energy.2020.118139

Open Access This chapter is licensed under the terms of the Creative Commons Attribution 4.0 International License (http://creativecommons.org/licenses/by/4.0/), which permits use, sharing, adaptation, distribution and reproduction in any medium or format, as long as you give appropriate credit to the original author(s) and the source, provide a link to the Creative Commons license and indicate if changes were made.

The images or other third party material in this chapter are included in the chapter's Creative Commons license, unless indicated otherwise in a credit line to the material. If material is not included in the chapter's Creative Commons license and your intended use is not permitted by statutory regulation or exceeds the permitted use, you will need to obtain permission directly from the copyright holder.

Natural Circulation and Other Measures to Ensure Heating Supply to Buildings Connected to District Heating in the Event of Electrical Grid Blackout

Merilin Nurme, Dabrel Prits, Karl-Villem Võsa, Andrei Dedov, and Anna Volkova

Abstract The increasing frequency of power outages, caused by severe weather events and the synchronization of the Baltic electricity grid with continental Europe, highlights the need for energy supply security. This is especially critical for residential buildings dependent on district heating (DH), which serves approximately 70% of apartment buildings in Estonia and over 80% in several Nordic cities. While cogeneration plants could provide DH network with heat by switching to "island mode" during blackouts, residential buildings typically lack backup generators, posing risks to uninterrupted heat supply. Research from Sweden found that natural circulation will occur in buildings' heating systems, allowing limited heat distribution during power outages. The present study aims to validate and assess the reproducibility of these findings under Estonian building conditions. Two field tests were conducted based on the hypothesis that natural circulation will occur if the temperature difference between the secondary circuit's supply and return water is sufficient. Results show that natural circulation can occur if primary valves remain open and sufficient temperature differences exist. Tests show other measures are needed to transfer heat from DH network to buildings during electrical grid blackout. This research proposes practical recommendations to support DH utilities and building owners in enhancing system resilience during blackouts.

Keywords District heating · Natural circulation · Space heating · Electrical grid blackout

---

M. Nurme (✉) · A. Dedov · A. Volkova
Department of Energy Science, Tallinn University of Technology, Tallinn, Estonia
e-mail: menurm@taltech.ee

K.-V. Võsa
Department of Civil Engineering and Architecture, Nearly Zero Energy Buildings Research Group, Tallinn University of Technology, Tallinn, Estonia

D. Prits
AS Utilitas Tallinn Soojus, Tallinn, Estonia

© The Author(s) 2026
D. Vanhoudt (ed.), *Proceedings of the 19th International Symposium on District Heating and Cooling*, Lecture Notes in Networks and Systems 1700,
https://doi.org/10.1007/978-3-032-09844-3_7

1 Introduction

Widespread power outages are becoming an increasingly common phenomenon. China has experienced severe power outages, primarily driven by significant increase in energy demand severe weather events [1]. Similar occurrences of increased power outages have been recorded in the Czech Republic, where the fault in the distribution center caused major disruption to public transport and urban infrastructures [2]. As most of the vital functions of modern society depend on the availability of electricity, including the heating systems of buildings, ensuring the security of energy supply has become a critical priority.

This issue is particularly relevant for residential buildings, as approximately 70% of apartment buildings in Tallinn, Estonia rely on district heating (DH), and in several Nordic cities, the figure exceeds 80% [3, 4]. Consequently, increasing emphasis is being placed on strategies that ensure stable energy supply during crises, while simultaneously minimizing environmental impacts [3, 5].

In this context, DH plays a vital role in achieving decarbonization targets. Centralized energy production enables easier regulation of fuel sources and more effective alignment of system parameters with climate objectives. The integration of renewable fuels into combined heat and power plants (CHP) has the potential to reduce greenhouse gas emissions by 16–70% [6, 7].

Given that climate policies strongly support the wider deployment of DH as a low-carbon solution, it is essential to acknowledge that the functionality of DH systems remains inherently dependent on a stable electricity supply. As power outages become more frequent, it is increasingly important to evaluate the performance and resilience of DH systems under such conditions.

Research from Sweden showed that, during power outages, natural circulation enabled buildings connected to district heating systems to recover 40–80% of their original heat supply within 3–4 h, with some cases achieving up to 90%. This level of residual heating may be sufficient to maintain acceptable indoor temperatures for several days. Based on these findings, the study emphasized the critical importance of securing back-up power for the operation of DH distribution networks, as their continued functionality is a prerequisite for enabling natural circulation in connected buildings [3].

In the case of Tallinn, the back-up power for the operation of DH distribution network is secured. It also needs to be considered that CHPs need large enough consumption from DH network to operate without auxiliary cooling system. This study investigates the feasibility of natural circulation for heat transfer in pump-operated heating systems during building-level power outages in district heating networks. Initial analyses indicated that the occurrence of natural circulation had low probability under the specified system parameters. Two experimental tests were carried out on typical five-story apartment building and a typical two-story kindergarten in Tallinn, selected based on representative statistical criteria. These tests were motivated by prior Swedish study suggesting that natural circulation may still be feasible in practice. In addition, the study explores alternative solutions for the operation of the heating substation under blackout conditions.

2 Methodology

2.1 Theoretical Basis of Natural Circulation

Natural circulation occurs when the pressure generated by natural convection exceeds the hydraulic resistance of the buildings internal pipeline system. When the water in the circuit is heated, its density decreases, causing it to expand and rise within the system. Conversely, as the water cools in the system, its density increases, promoting it to descend back towards the heat source. The natural circulation pressure p_g is the difference between the pressures in supply p_s and return p_r flow and p_g can be found with following equation [8].

$$p_g = p_s - p_r = h \times (\rho_r - \rho_s) \times g \quad (1)$$

where p_g is the natural circulation pressure; h is the height between the secondary circuit supply and return pipe; ρ_r the water density in return pipe; ρ_s the water density in supply pipe; g stands for the standard gravity.

For natural circulation to occur in pump-driven heating system it is essential for the natural circulation pressure to exceed the pressure losses in the building's internal pipeline system. Pressure losses in the building's internal pipeline are determined by Darcy-Weisbach equation [8].

$$\Delta p_i = \frac{\lambda_i}{d_{s,i}} l_i \frac{\rho \times w_i^2}{2} + \sum_{i=1}^{n} \varsigma_i \times \frac{\rho \times w_i^2}{2} \quad (2)$$

where Δp_i is pressure losses in a certain section of the pipeline; λ is hydraulic friction coefficient; $d_{s,i}$ is pipeline inner diameter; w is velocity of the heated water; l the length of pipeline section; ς is resistance coefficient; ρ is the water density in the pipeline section.

2.2 Site Requirements and Test Execution

During the field tests, the supply and return water temperatures in the secondary circuit, as well as the indoor air temperature, were monitored. Temperature changes in the secondary circuit were recorded using surface-mounted sensors (PeakTech Clamp Temperature Probe, model P TF-25, Type K, –20 to +200 °C) attached to the

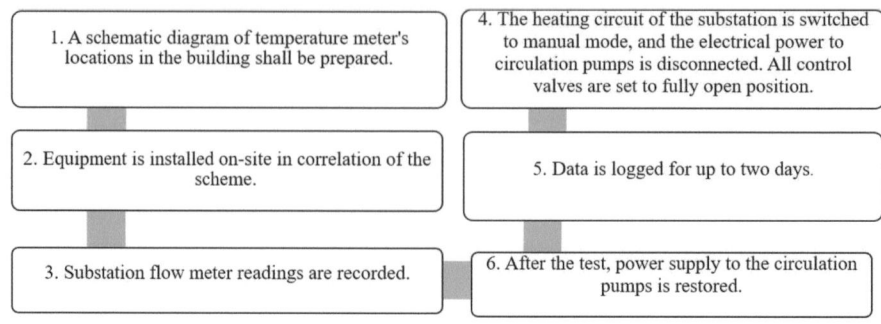

Fig. 1 Overview of the field test procedure

internal piping system. The volumetric flow rate was measured with an ultrasonic flow meter (Wall Mount TUF-2000B). Surface-mounted temperature sensors, insulated with foam plastic, were specifically used to measure the supply temperature of the heat carrier in the secondary circuit.

The following requirements were established for the field tests:

- The optimal outdoor air temperature range was 0 to –10 °C, based on the assumption that colder conditions would accelerate cooling of water in the secondary circuit's distribution piping, thereby increasing the temperature differential needed to overcome hydraulic resistance;
- Access to the building's heating supply risers and secondary circuit radiators is available, and control valves can be fully opened;
- Substation flow meter data is remotely accessible, and the building substation's heating circuit allows for manual control;
- All equipment used were adjusted according to the parameters of the building heating system and the test conditions;
- For accurate interpretation and validation of the results obtained from the test buildings, comparison sites with comparable building parameters are required.

The following steps were taken to conduct the field tests (see Fig. 1).

3 Results

3.1 Hydraulic Calculations of the Buildings' Internal Pipeline System

Prior to the disconnection of power to the circulation pump, the supply and return water temperatures in the secondary circuit were 46 and 38 °C, respectively, in the apartment building, and 42 and 38 °C in the kindergarten. The supply water flow rate was 8.17 kg/s in the apartment building and 12 kg/s in the kindergarten.

Table 1 Results of hydraulic calculations for the internal pipeline systems of the field test site buildings (see Eqs. (1) and (2))

	Natural circulation pressure p_g (kPa)	Pressure losses in the internal pipeline system of the buildings (kPa)
Field test site no. 1: Apartment building	31,29	0,275
Field test site no. 2: Kindergarten	20,83	0,106

As indicated by the calculations (see Table 1), the occurrence probability of natural circulation is low, as the pressure losses exceed the driving pressure generated by natural convection.

3.2 Field Test No. 1: Apartment Building

The field test in the apartment building (see Fig. 2) was conducted on 12 December 2024, between 09:45 and 17:00 in Tallinn. During the experiment, the outdoor air temperature ranged from 0 °C to –2 °C. In the primary circuit, the control valve of the space heating heat exchanger was fully opened, while the control valve for the domestic hot water (DHW) HEX remained in the same position as prior to the power outage, ensuring that residents had access to hot water throughout the test period.

Temperature measurement points were installed only on the supply lines of radiators and on risers located in the corridors. In the other stairwells, measurements were taken both at the basement level and in the corridor (see Fig. 3).

After the power supply to the circulation pump was disconnected, the supply-side flow rate decreased from 8.17 m³/h to 0.65 m³/h, corresponding to approximately 7.8% of the initial flow (see Fig. 4). Simultaneously, the supply temperature in the

Fig. 2 Field test location no. 1: The apartment building [9]

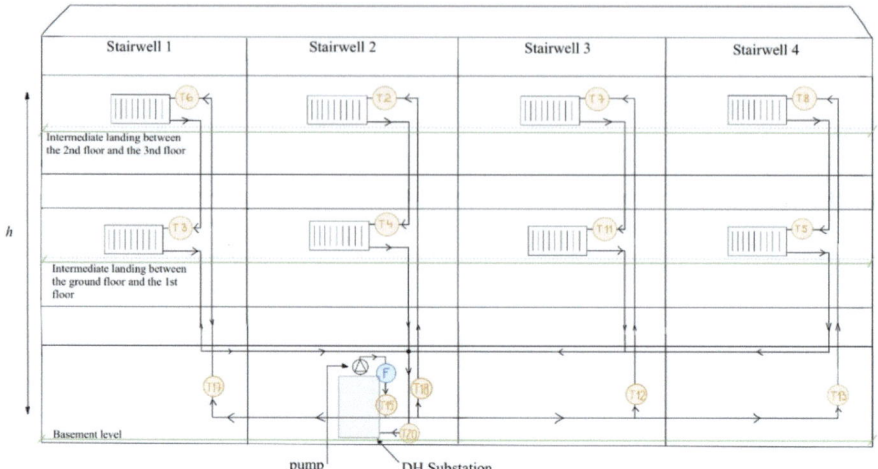

Fig. 3 A schematic drawing of measurement points in the apartment building. The image indicates the *T2–T20* as temperature measurement points located in the apartment building; *F* as volumetric flow meter; *h* as the height difference between heat exchanger and radiator

secondary circuit increased to 73.5 °C, while the return temperature dropped to 25 °C. This indicates a significant increase in temperature differential, despite the reduced flow.

Although heat consumption generally correlates with flow rate, the system exhibited irregular fluctuations after reaching a new thermal equilibrium. These variations can be partly attributed to the limitations of the flow meter device: any value below

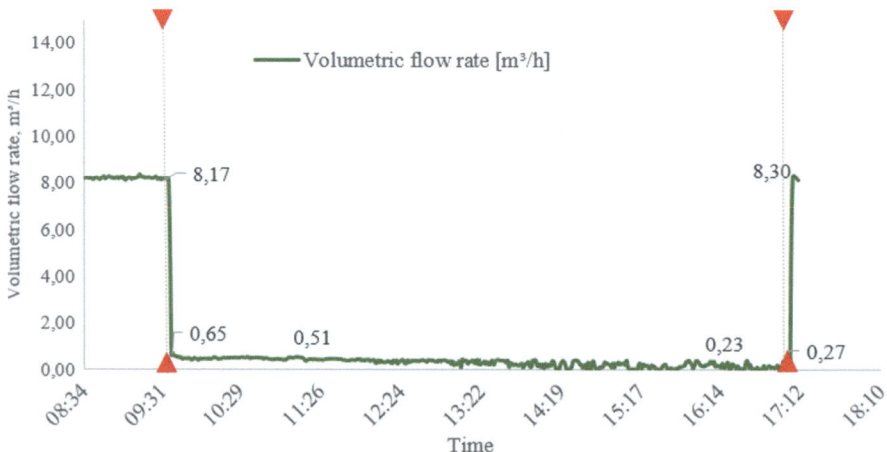

Fig. 4 Volumetric flow rate in the apartment's secondary supply manifold

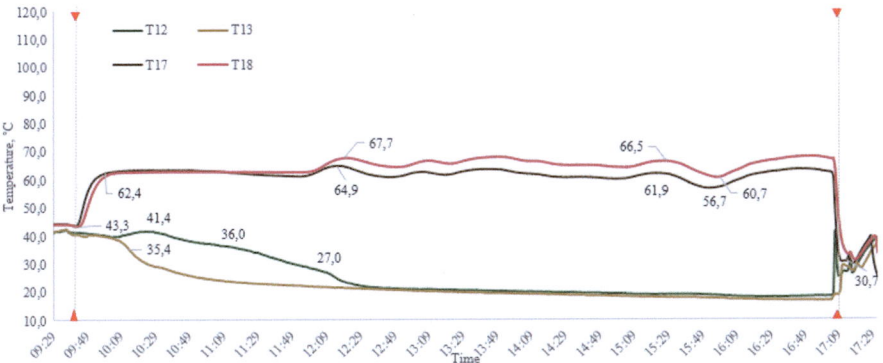

Fig. 5 The temperatures measured in the apartment's basement level

0.13 m³/h is recorded as zero, which explains the discontinuities observed in the flow data during the period preceding the restoration of power.

Measurement devices were installed on the supply risers of the basement-level distribution piping. Prior to the power outage, temperatures in the risers ranged from 39.8 to 46 °C (measurement points T12 and T13). In the first and second stairwells (measurement points T17 and T18), temperatures increased from 40 to 68.9 °C, indicating effective heat transfer from the substation to the risers (see Fig. 5). In the third stairwell, the riser temperature rose slightly from 40 to 41.4 °C (measurement point T12), but subsequently dropped to 17.8 °C, suggesting that hydraulic resistance hindered sustained flow. At the end of the building, temperatures declined from 39.8 to 17.8 °C, indicating a lack of natural circulation in the outermost risers, as evidenced by the consistent cooling trend.

Prior to the test, the radiator supply riser temperatures in the stairwells ranged from 36.9 to 40.5 °C (see Fig. 6). Following the disconnection of the circulation pump, the temperatures dropped by up to 4 °C. However, within 15 min, a slight increase of up to 1 °C was observed on all risers (measurement points T3, T5, and T11).

The minimal temperature rise between the first and second floors suggests that the heating water was gradually cooling. In a single-pipe heating system, heat is initially delivered to the upper floors before descending through the radiators. The observed increase in corridor radiator temperatures indicates that warmer water moved into cooler sections of the circuit. A brief episode of natural circulation likely occurred in the secondary circuit, as evidenced by the temperature rise in the basement distribution system. However, due to hydraulic resistance, this flow was insufficient to deliver heat to the radiators on the upper floors via the risers.

When comparing the heat consumption capacity of the apartment building and the reference building during the experiment, a sharp decline is evident (see Fig. 7). Following the interruption of electricity supply, the apartment building's heat consumption capacity decreased from 148 kW to 17,5 kW on primary circuit heat meter. A partial recovery was observed, with the capacity increasing up to

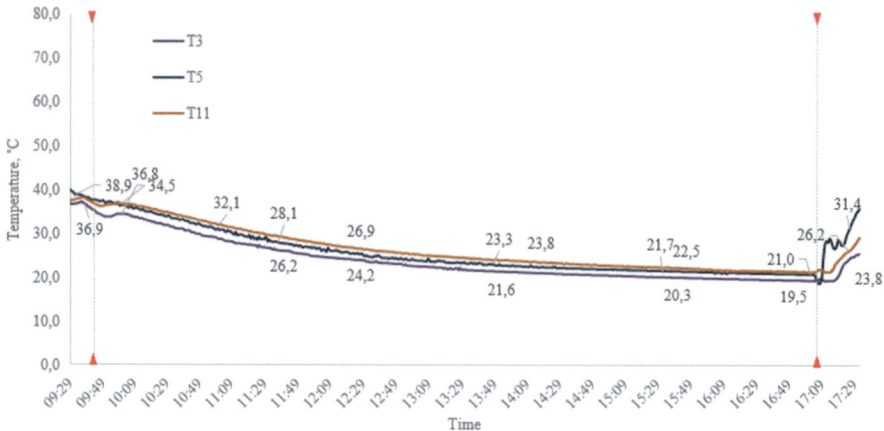

Fig. 6 The temperature measured in the apartment's intermediate landing between the ground floor and 1st floor

24 kW; however, normal levels were restored only after the electricity supply to the circulation pumps was re-established. The apartment building's heat consumption remained at approximately 16% of the pre-interruption level throughout the outage; however, this figure also accounts for the continued partial operation of the domestic hot water system, which remained open during the test.

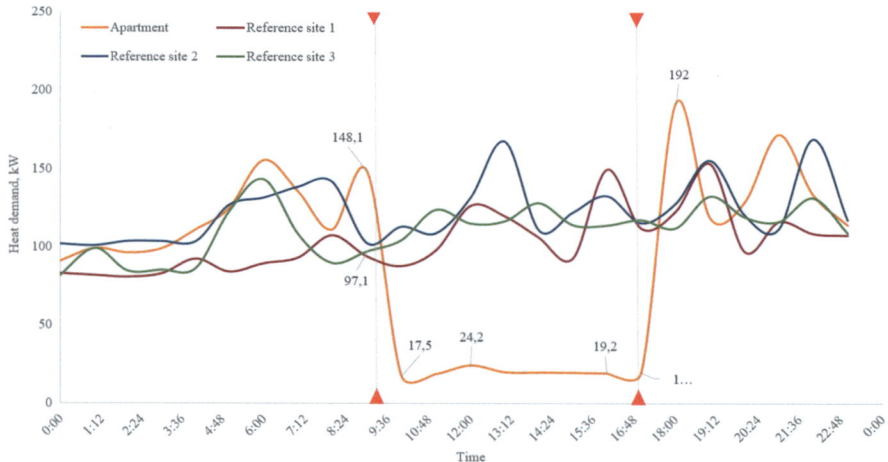

Fig. 7 Heat demand comparison between the apartment building and reference sites. To ensure comparability, control sites were selected based on spatial proximity and building-type similarity

3.3 Field Test No. 2: Kindergarten: Analysis of Results and Conclusions

The measurement at the kindergarten (see Fig. 8) was carried out from 22 November 2024 at 20:00 until 24 November 2024 at 16:00 in Tallinn. During this period, the circulation pump's power supply was disconnected on 22 November 2024 at 20:07 and reactivated on 24 November 2024 at 12:00. Throughout the experiment, the outdoor air temperature ranged between −3 and −1 °C. The control valve on the primary circuit of the space heating heat exchanger was fully opened, while the control valve of the DHW heat exchanger remained in the same position as prior to the power outage.

In addition to temperature, the flow rate in the supply line of the secondary circuit was measured in the kindergarten. The measurement device was installed immediately after the heat exchanger, on the supply side of the distribution piping. Based on the results, no natural circulation occurred in the secondary circuit following the disconnection of power to the circulation pump, as the flow rate dropped immediately to zero and only resumed after the restoration of the electricity supply (see Fig. 9).

Fig. 8 Field test location no. 2: The kindergarten facility [9]

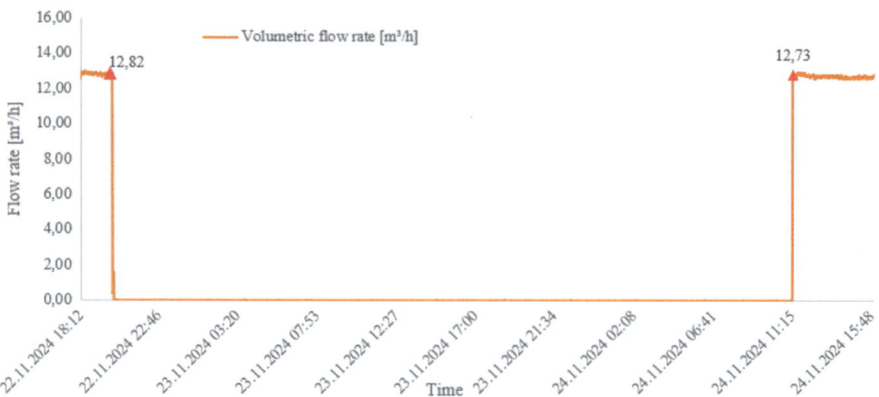

Fig. 9 Volumetric flow rate in the kindergarten's secondary supply manifold

3.4 Other Measures for Customer Buildings Connected to DH

To ensure the continuous operation of district heating (DH) networks, it is essential that heat supply to buildings remains functional during electricity outages. Based on field experiments conducted in an apartment building and a kindergarten, the following recommendations are made for DH consumers:

- In the absence of an emergency power supply, provision should be made for its installation—namely, ensuring the possibility of connecting a generator to the building's or heating substation's electrical system;
- Heating systems and substations that are outdated or inefficient should be upgraded or replaced with more energy-efficient equipment;
- Building renovation should aim to improve thermal insulation and minimize heat losses.

Regarding implementation feasibility, in Estonia, the cost of installing a generator connection for a typical apartment building's circulation pump can remain under €1000 (see Table 2). Since this setup typically requires linking only to the heating substation's electrical cabinet, not the entire building's electrical system, it is a relatively simple and cost-effective solution for most apartment buildings.

Table 2 Estimated power requirements and backup supply options for residential heating substations

Substation power consumption	Required continuous output	Recommended backup power source	Estimated cost
Below 1500 W (e.g. 700 W)	≥1800 W	Inverter generator (e.g. 1700 W)	Generator: from €400–600 (estimate)
		12 V car battery + inverter (1500 W)	Inverter (12 V, 1500 W) ≈ €330 + car battery
1500–2500 W	≥2800 W	Inverter generator (≥2800 W)	Generator: from €400–600 (estimate)

4 Conclusion

With increasing power outages driven by climate events and grid disruptions, ensuring the resilience of district heating (DH) systems is essential—particularly in regions like Estonia, where a majority of residential buildings depend on DH for space heating. Motivated by a Swedish study that demonstrated the potential of natural circulation to restore 40–90% of heat supply within 3–4 h during power outages, this study set out to assess the reproducibility of those findings in typical Estonian building contexts.

Based on this precedent, hypothesis proposed that natural circulation could maintain partial heat transfer if sufficient temperature differentials existed in the secondary circuit. However, hydraulic calculations and field experiments in a five-story apartment building and a two-story kindergarten revealed that internal pressure losses generally exceeded the driving pressure from natural convection. Only limited, short-term natural circulation was observed in the apartment building, while no circulation occurred in the kindergarten. Heat recovery peaked at just 16% of pre-outage consumption, far below the levels reported in the Swedish case.

These results suggest that natural circulation alone does not ensure reliable heat delivery in most DH-connected buildings without system adaptations. To address this, practical recommendations include enabling generator connection to the heating substation's electrical cabinet, upgrading outdated equipment, and improving building insulation. Importantly, the cost of implementing backup power solutions remains relatively low, under €1000 for a typical apartment building, making them both feasible and impactful.

While natural circulation may offer limited thermal buffering, maintaining heat supply during blackouts requires proactive measures. Low-cost investments in backup power and efficiency improvements are essential to strengthen the resilience of DH systems in the face of growing energy security challenges. Although this study was conducted in Estonia, the findings and recommendations are applicable to other countries with similar building typologies, heating systems, and dependency on district heating.

Acknowledgements Utilitas: test and data provision. This research was co-funded by the European Union and the Estonian Research Council via projects TEM-TA78 and PRG2701.

References

1. Hoskins P (2021) China power cuts: What is causing the country's blackouts? BBC News. [Online]. https://www.bbc.com/news/business-58733193. [Accessed: May 28, 2025]
2. Lazarová D (2024) Power outages complicating evacuation efforts. Radio Prague International. [Online]. https://english.radio.cz/power-outages-complicating-evacuation-efforts-8828836. [Accessed: May 28, 2025]
3. Lauenburg P, Johansson P-O, Wollerstrand J (2010) District heating in case of power failure. Appl Energy 87(4):1176–1186. https://doi.org/10.1016/j.apenergy.2009.08.018
4. Utilitas, "District heating," Utilitas. [Online]. https://utilitas.ee/kaugkute-ja-kaugjahutus/kaugkute/. [Accessed: May 28, 2025]
5. Nik VM, Sasic Kalagasidis A (2013) Impact study of the climate change on the energy performance of the building stock in Stockholm considering four climate uncertainties. Build Environ, 60:291–304. https://doi.org/10.1016/j.buildenv.2012.11.005
6. Lake A, Rezaie B, Beyerlein S (2017) Review of district heating and cooling systems for a sustainable future. Renew Sustain Energy Rev 67:417–425. https://doi.org/10.1016/j.rser.2016.09.061
7. Malcher X, Tenorio-Rodriguez FC, Finkbeiner M, Gonzalez-Salazar M (2025) Decarbonization of district heating: A systematic review of carbon footprint and key mitigation strategies. Renew Sustain Energy Rev 215:115602. https://doi.org/10.1016/j.rser.2025.115602
8. Kõiv TA, Rant A (2013) Hoonete küte. TTÜ Kirjastus, Tallinn, Estonia
9. Estonian Land and Spatial Development Board, "Fotoladu—Orthophoto viewer." [Online]. https://fotoladu.maaamet.ee/?basemap=hybriidk&zlevel=13,24.66739,59.39746&overlay=avaleht. [Accessed: May 28, 2025]

Open Access This chapter is licensed under the terms of the Creative Commons Attribution 4.0 International License (http://creativecommons.org/licenses/by/4.0/), which permits use, sharing, adaptation, distribution and reproduction in any medium or format, as long as you give appropriate credit to the original author(s) and the source, provide a link to the Creative Commons license and indicate if changes were made.

The images or other third party material in this chapter are included in the chapter's Creative Commons license, unless indicated otherwise in a credit line to the material. If material is not included in the chapter's Creative Commons license and your intended use is not permitted by statutory regulation or exceeds the permitted use, you will need to obtain permission directly from the copyright holder.

Feasibility Assessment Tool for District Heating and Cooling (FAST DHC): A Simple Decision Support Tool for the Techno-Economic Evaluation of DHC Networks

Henrique Lagoeiro, Nicolas Marx, Alessandro Maccarini, Oddgeir Gudmundsson, Catarina Marques, Ralf-Roman Schmidt, and Graeme Maidment

Abstract The Feasibility Assessment Tool for District Heating and Cooling (FAST DHC) project, funded through the IEA DHC Annex XIV, aims to develop and demonstrate a simple, freely available, web-based decision support tool for the techno-economic performance evaluation of 4th generation district heating (4GDH) and thermal source networks (TSNs), whilst also enabling their comparison to individual heating and cooling (H&C) solutions. The TSN concept is in its early stage of development and there is a lack of understanding of its relative merits against traditional DHC concepts, i.e. how do 4G and TSN systems compare and what are their competitive advantages to individual H&C systems. The FAST DHC tool will enable users (e.g. local authorities, designers and energy planners) to perform early-stage feasibility studies and easily compare the potential benefits of the latest DHC typologies, providing greater clarity on how/where each system may be best applied. The aim of the tool is to assist the development of the DHC sector by equipping users with a reliable initial estimate of proper system setup, therefore maximising benefits to DHC developers, operators and end users. This paper introduces the FAST DHC tool, describing its design and functionalities, and provides the results from a case study where the FAST DHC tool has been compared to a commercial tool for the techno-economic modelling of DHC systems.

H. Lagoeiro (✉) · C. Marques · G. Maidment
London South Bank University, London, UK
e-mail: roscoeh2@lsbu.ac.uk

N. Marx · R.-R. Schmidt
Austrian Institute of Technology, Vienna, Austria

A. Maccarini
Aalborg University, Aalborg, Denmark

O. Gudmundsson
Danfoss A/S Climate Solutions, Nordborg, Denmark

Keywords District Heating and Cooling (DHC) · Techno-economic evaluation · Planning tools · 4th generation (4GDH) · Thermal Source Networks (TSNs)

1 Introduction

1.1 Context

District heating (DH) has been traditionally deployed as an efficient technology for meeting the thermal needs of cities, particularly due to its significant economies of scale. The first generations of DH are defined by high operating temperatures and dominance of fossil-based heat sources. In the last decade, a modern DH concept has emerged which encourages operation at much lower temperatures, increasing energy efficiency and unlocking the potential for decarbonisation with renewable and waste heat sources. This transition has been conceptualised as the 4th generation of DH (4GDH), whose networks are characterised by flow temperatures below 70 °C [1]. The most common typology of 4GDH systems provide only heating (i.e. no cooling supply) via central generating plants, with operating temperatures close to the upper limit that defines 4GDH (60–70 °C). In recent years, the concept of thermal source networks (TSNs), a new subclass of 4GDH, has been introduced. Sometimes referred to as the "5th generation", these networks are characterised by the integration of consumer-side heat pumps (HPs), which effectively removes restrictions of system operating temperatures, as they can boost the temperature to the internal requirements of connected buildings. They further enable heating and cooling (H&C) to be supplied within one network.

The differences in these typologies of low-temperature district heating and cooling (DHC) networks have led to a fundamental debate within the international DHC community over the role and applicability of TSNs, particularly as the commonly used term "5th generation" might lead to the perception that TSNs are a progression from the 4GDH, which is not always the case [2]. This was demonstrated in [3], where for a single-source system without cooling, 4GDH was shown to be economically superior to a TSN configuration. On the other hand, TSNs could be a viable alternative in mild climates with heating and cooling demands of similar magnitudes, particularly where traditional DH systems are not present [4].

1.2 The FAST DHC Project

This ongoing debate over the merits and challenges of traditional 4GDH systems and TSNs was the main motivation for the FAST DHC project. There is uncertainty over the role and applicability of TSNs, and in which cases they can outperform a 4GDH network or individual H&C systems. Typically, the selection of a certain DHC typology demands a case-by-case detailed assessment to identify the best system design, and there is no "one-size-fits-all" solution. This is a data-intensive process that often requires the use of commercial tools, which typically require information (e.g. hourly demand data) that might not be available in early planning stages. Most existing tools either require granular data (e.g. EnergyPLAN, energyPRO) or are unable to provide a direct comparison between different DHC typologies and individual solutions (e.g. EMB3Rs, Thermos, HotMaps).

Conversely, The FAST DHC project is aimed at supporting early-stage decision making with a freely available web-based tool for the techno-economic performance evaluation of 4GDH and TSNs, and their comparison to individual H&C solutions. The tool should give users an indication as to which H&C system represents the best potential investment under the assumed boundary conditions, being particularly useful for end-users with limited resources or expertise to perform more detailed analyses. This includes municipalities and consultants conducting first-phase assessments, where hourly data is not yet available. Therefore, the novelty of FAST DHC lies in the way that it can provide a pre-feasibility comparison with very little input from users.

This paper introduces the FAST DHC tool by describing its techno-economic methodological framework and its database of technological components. Furthermore, results from a case study are presented and used to indicate the validity of FAST DHC when compared to a feasibility analysis carried out with a commercial software tool.

2 Technical Modelling Approach

2.1 FAST DHC Tool Concept

The FAST DHC tool is designed to be a modular, web-based feasibility assessment application that allows users to quickly evaluate different DHC configurations and how they compare to individual H&C solutions. It combines techno-economic modelling with minimal user inputs and relies heavily on pre-populated datasets and look-up tables, based on publicly available sources, such as Eurostat and published data from other research projects related to heating and cooling.

The tool's front-end and back-end interact dynamically: the back-end relies on Python and Dash callbacks to update calculation modules and visual outputs in real time. The front-end allows for iterative user interaction. The core logic is structured around a parameter sheet, which serves as a comprehensive, editable repository for all techno-economic defaults and lookup structures. By doing so, the FAST DHC tool enables robust, early-stage feasibility analysis while remaining accessible to users across various expertise levels. In summary, the FAST DHC tool balances simplicity with reliability by:

- Using empirical equations and previous datasets from trusted sources instead of simulations;
- Providing seasonal granularity instead of hourly resolution;
- Allowing for traceable, transparent calculations in the graphical user interface (GUI);
- Containing a flexible, modular structure that enables updates to data tables and assumptions by users.

2.2 Description of System Configurations

The FAST DHC tool will allow users to input a few relevant parameters, specify demands and supply options, and this information will be used to estimate the levelised costs of energy (LCOE), including both heating and cooling, for each of the following system configurations:

- Traditional 4th Generation District Heating (4GDH) combined with individual cooling;
- Thermal Source Networks (TSNs) enabling bidirectional heat flows and shared infrastructure;
- Individual Heating & Cooling (H&C) solutions using decentralised air-source heat pumps.

These system set-ups are illustrated in Fig. 1. The levelised cost estimations are based on a database of technological components (e.g. generation plants, storage elements, pipework, substations) that provides the buildings blocks for all H&C system configurations considered. Input data is simplified to four characteristic time slots (winter, spring, summer and autumn), which reduces computational effort whilst also considering the seasonal mismatch between H&C demands and supply. Each system configuration can optionally incorporate thermal storage either at a building level, for domestic hot water (DHW), or on a network level, which enables flexible heat pump operation based on electricity market prices. Furthermore, the user can select different heating emitters for the connected buildings (radiator versus underfloor heating), resulting in different temperatures, centralised or decentralised supply set-ups, and booster components.

2.3 Demand-Side Modelling

The FAST DHC tool has an inbuilt demand estimation that is based on data provided by the user on the number of buildings connected, their types (single family, multi-family, or non-residential) as well as their age (old versus new stock) and emitters (radiator versus underfloor heating). This approach reduces the input burden on the user and relies on a set of look-up tables containing the following data:

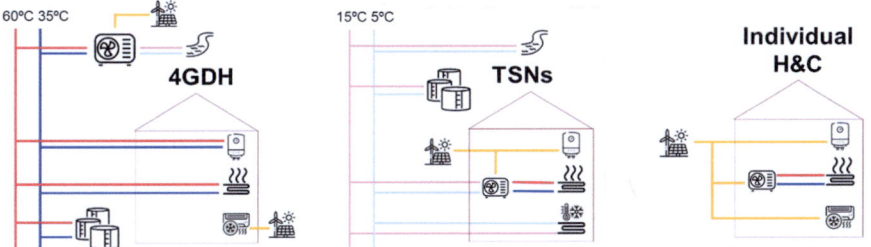

Fig. 1 Different H&C system configurations considered in the FAST DHC tool

- Heating degree days (HDD) [5] and cooling degree days (CDD) [6] by country and climate region;
- Building archetypes [7] and their average floor areas [8] for different countries;
- Default full-load hours, temperature levels, and thermal loads for space heating [9], cooling [10] and DHW [11].

Seasonal energy profiles are automatically derived from country-specific HDD/CDD distributions and partitioned across the four seasons: winter, spring, summer and autumn. Users may override these defaults, though they are aligned with datasets such as Eurostat [12] and Hotmaps [13]. Heating demand is disaggregated by building type and age (old versus new stock) and further by emitter type (radiator versus underfloor heating). Cooling and DHW demands follow analogous rules. Building-specific energy demands are calculated by combining floor area (m^2), load intensities (W/m^2), and full load hours, all customisable by the user. The central data structure consists of 48 seasonal demand values per simulation case (3 building types × 2 age groups × 4 seasons × 2 emitter types), which are organised in a look-up structure and visualised as an input matrix in the GUI.

2.4 Supply-Side Modelling

The supply model dynamically reacts to the calculated seasonal demands. Central heat pump (HP) capacities for 4GDH and TSNs are sized based on seasonal peak load and user-defined coverage. Backup capacity is calculated if this combination does not meet peak demand. Each heat source is characterised by average seasonal temperatures, capacity constraints, and associated costs, all of which are estimated using publicly available data and can be adapted by the user. Combustion heat sources are modelled with a fixed efficiency, and heat pumps are modelled with a calculated coefficient of performance (COP), as shown in Eq. 1. This is based on the seasonal variation of source and network supply temperatures, expressed in Kelvin (K), with the latter being fixed for 4GDH (60 °C) and TSNs (15 °C), considering a Carnot efficiency of 50%.

$$COP = 0.5 \bullet \frac{T_{supply}}{T_{supply} - T_{source}} \quad (1)$$

2.5 Calculating Network Length

The effective width methodology is used to determine the length of the DHC network in applicable scenarios. This methodology can be used to estimate network size based on the total land area covered by the network, which is estimated from the typical building densities for urban, suburban and rural areas. Although there is some degree of uncertainty in this approach, Sánchez-García et al. reported a good correlation between network length and building density based on data from DH companies in Denmark (see Sect. 3.1.2 in [14]). This is reflected in Eq. 2, which was incorporated into the FAST DHC tool as it provides the best fit amongst the analysed equations for calculating network length (L, in metres per hectare) from the number of buildings per hectare (N).

$$L = \frac{213.191}{1 + 19.111 \bullet N^{-1.532}} [\frac{m}{ha}] \quad (2)$$

2.6 Database of Technological Components

The FAST DHC database comprises 13 distinct technological components relevant to H&C systems in urban energy networks. These components represent a broad range of technologies, including but not limited to, heat pumps, boilers, DH substations, and thermal storage units. Where applicable, components are further categorised by attributes such as size, location, and building type, ensuring that the database can support scenario-specific customisation and modular system design.

The database was implemented in JavaScript Object Notation (JSON) format, selected for its flexibility and compatibility with modern web-based tools. JSON's support for nested structures allows for the efficient representation of hierarchical data, such as arrays and objects, which is especially valuable in modelling complex energy systems. Moreover, as a native format in JavaScript environments, JSON enables straightforward integration with front-end and back-end tools for parsing, manipulation, and visualisation.

The primary data sources for the database are the Danish Technology Catalogues, developed by the Danish Energy Agency in collaboration with affiliated research institutions [15]. These catalogues provide detailed techno-economic performance data for a wide range of energy technologies and are routinely used for policy planning and energy projections in Denmark. Specifically, three catalogues were referenced,

Table 1 Summary of the components in the FAST DHC database

Category	Component	Attributes
Plant	Biomass boiler	Size: small, medium, large
	Heat pump (air source)	Size: small, medium, large
	Heat pump (excess/waste heat)	Size: small, medium, large
	Heat pump (sea water)	Size: medium
	Heat pump (geothermal)	Size: medium, large Temperatures: 40–80°C, 35–70°C Depth: 1200, 2000 m
	Electric boiler	Size: small, large
	Heat exchanger	Location: rural, suburban, city
	Gas boiler	Size: medium
Distribution	Piping network (4G or TSN)	Location: rural, suburban, city
Building level	Heat exchanger (for indirect connections)	Building type: single or multi-family Age: new, existing
	Heat pump (air-to-water)	Building type: single or multi-family Age: new, existing
	Heat pump (ground-to-water)	Building type: single or multi-family Age: new, existing
	Heat pump/ air conditioner (air-to-air)	Building type: single family Age: new, existing

relating to the generation of electricity and district heating, individual heating plants and transport of energy. Each component in the database includes information on investment costs, operational parameters, lifetimes, efficiency, and environmental indicators, enabling their direct use in techno-economic assessments. A summary of the components in the database is provided in Table 1.

3 Economic Modelling Approach

3.1 Cost Estimation Methodology

When estimating the costs for heating and cooling of buildings, which are ongoing demands over the lifetime of a building, there are a few important factors that must be considered: initial investment (CAPEX), cost of capital (interest rates), operational and maintenance cost (OPEX), technical lifetime, reinvestments and annual heating and cooling demands. These parameters can be grouped into investment and operational elements in the economic model, as shown in Fig. 2.

This simple economic model can be extended from the basic form, which would represent individual building level solution, to a more complex model representing

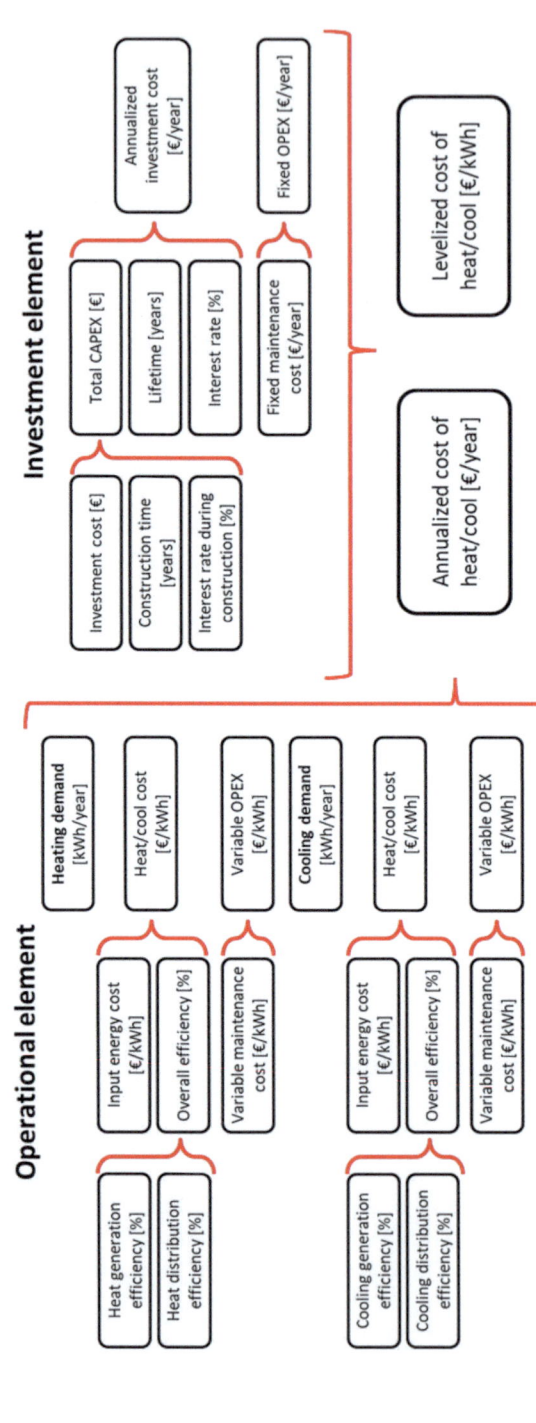

Fig. 2 Economic model for estimating levelised and annual costs of heating and cooling

district energy systems, by adding additional investment and operational elements. This modular functionality of the economic model enables quick assessment of the cost of heating and cooling, either through LCOE or total annualised cost for the system configurations analysed.

The main challenge with using the economic model is to acquire location specific input data. For individual building solutions, these can normally be acquired from the local markets being considered. For infrastructure-based solutions, like district energy systems, acquiring necessary technology data can be a challenge. However, these can be found in various technology catalogues, as described in Sect. 2.6.

3.2 Effective Use of Technology Data

The FAST DHC tool is designed to provide a quick pre-feasibility assessment of H&C configurations without the need for extensive data collection. As such, it becomes necessary to rely on published technology catalogues, which provide cost and efficiency data for relevant technologies at a specific year with a certain capacity. This poses a modelling challenge, as the data in catalogues might not be aligned with the requirements of the user. Furthermore, due to the scarce availability of technology cost data in different countries, it may be necessary to rely on a foreign catalogue, which then requires the cost data to be transferred to the country of interest and potentially updated to account for inflation from the date of the cost data to the date of the analysis. These three challenges, (a) discrete capacities of technologies, (b) transfer of cost data between countries and (c) updating costs due to inflation, need to be addressed to ensure the best estimates of LCOE.

For addressing (a), at least two data points are necessary, consisting of values for capacity and cost, which can be used for fitting an exponential curve, which enables estimating the cost of the given technology at different capacities. Figure 3a illustrates this approach. For addressing (b) and (c), it is possible to apply Purchasing Power Parity (PPP) factors, which compare costs internationally, for estimating the costs of comparable technologies between countries, and inflation datasets can be used for adjusting costs from the date of the technological datapoint to the date of the analysis. The PPP factors are derived from Eurostat's datasets for Machinery and equipment (A0501) and Civil engineering works (A050203), whereas inflation factors are obtained from Eurostat datasets for Manufacturing of machinery and equipment (STS_INPP_A: C28), for all technology equipment cost data, and Services of plumbers (PRC_HICP_AIND: CP04321), for all civil work [16]. The approach for transferring data between countries and updating them to the present value is illustrated in Fig. 3b.

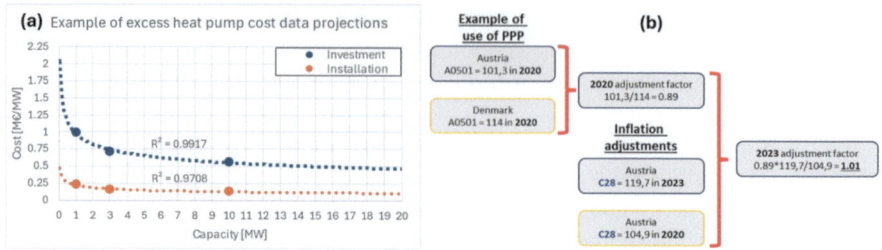

Fig. 3 Example of cost projections for varying capacities using power functions (**a**) and example of usage of PPPs and inflation adjustments datasets (**b**)

4 Tool Demonstration and Initial Validation

4.1 Initial Validation with a Case Study

A critical part of the FAST DHC project will involve testing the FAST DHC tool with diverse case studies related to the different system configurations described in Sect. 2.2. Four case studies that consist of feasibility assessments carried out with commercial software tools will be replicated on FAST DHC for validation purposes. These case studies are from the countries of the partner institutions (Austria, Denmark and the UK). At the time of writing, only the analysis of the first case study of a 4GDH system had been completed, and the results of the replication are described in this section.

4.2 Description of Case Study 1

The first case study used for validation consists of a feasibility study for a 4GDH network in Bristol, UK, operating with a flow temperature of 65 °C. The main heat source is a 750 kW$_{th}$ water-source heat pump recovering waste heat from a large supermarket store. The suggested design would entail recovering waste heat from the supermarket's remote refrigeration system, which rejects heat via air-cooled condensers mounted on the rooftop. In order to enable heat recovery, these condensers would have to be replaced with water-cooled variants that would be connected to an ambient loop, shown in green in Fig. 4. After upgrade, the heat would be supplied to a residential development consisting of 284 households spread across 8 multi-family buildings (5 new and 3 existing, all with radiator heating), which are located next to the supermarket, as can be observed in Fig. 4. The existing buildings are served by communal gas boilers which would be retained to meet peak loads, and 75 m^3 of thermal storage would be included in the energy centre to smooth the operation of the heat pump and balance supply with demand. More details on this case study can be found in [17].

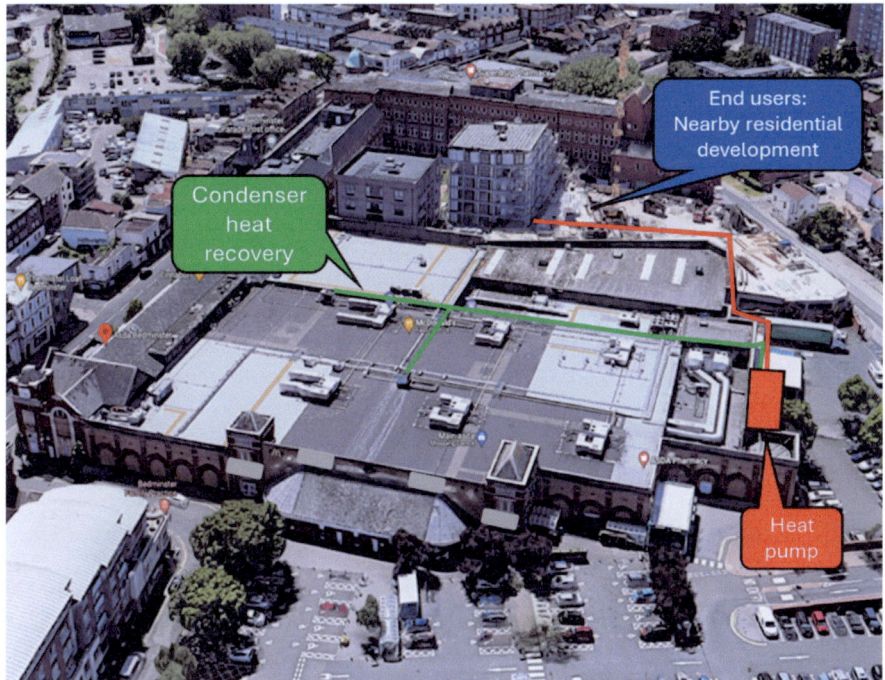

Fig. 4 Aerial view of the supermarket and residential development considered in case study 1

The case study techno-economic assessment was undertaken using energyPRO to estimate the OPEX, internal rate of return and carbon savings against a gas boiler counterfactual. The validation work involved expressing the cost figures from the case study analysis in the FAST DHC format, and energy costs also had to be adjusted as the case study used 2019 values, with an electricity price of €150/MWh, as opposed to the €350/MWh assumed in the tool for a non-domestic user (i.e. the heat network operator). The network supply temperatures within the tool were also changed to 65°C to match the design supply temperature of the case study.

4.3 Results and Discussion

The comparison of case study and FAST DHC results is provided in tabular (Table 2) and graphical (Fig. 5) formats. As can be observed, there is good agreement between the values estimated by the FAST DHC tool and the energyPRO case study, with most values within a 20% difference. Some discrepancies include capacities for the thermal store and the heat pump, as our estimation relies on a generic heating degree-days approach, impacting CAPEX. Heat recovery costs were also quite different; in FAST DHC, heat recovery follows generic estimates based on data provided by the

UK Government [18] and is included under OPEX as a levelised cost of capture, which is different from the case study, where investment costs for heat recovery equipment were calculated in detail and included under CAPEX. This difference impacted OPEX estimations significantly, but can be deemed reasonable as the generic approach adopted in FAST DHC accounts for longer connection distances than that of the case study, where the heat pump would be located at the heat recovery site.

As can be observed from Fig. 5 with greater clarity, the main discrepancy in estimates is linked to the costs for recovering heat from a central source. If the value from the case study is used in the FAST DHC tool, something that can be adjusted by the user, the annualised costs for the 4G network becomes €180,052 (2% higher

Table 2 Comparison of modelling results for FAST DHC and the case study

Item	FAST DHC	Case Study	Difference
Space heating demand (MWh/year)	1759	1867	− 6%
Hot water demand (MWh/year)	1053	1042	+ 1%
Cooling demand (MWh/year)	267	N/A	N/A
Heat pump capacity (kW)	640	750	− 15%
Thermal storage size (m^3)	68	75	− 9%
Network length (m)	80	80	0%
Total CAPEX (EUR)	€986,022	€1,050,953	− 6%
Total OPEX (EUR)	€145,549	€115,049	+ 27%
Annualised costs (EUR/year)	€205,244	€177,046	+ 16%
Levelised costs of energy (EUR/MWh)	€73	€61	+ 20%

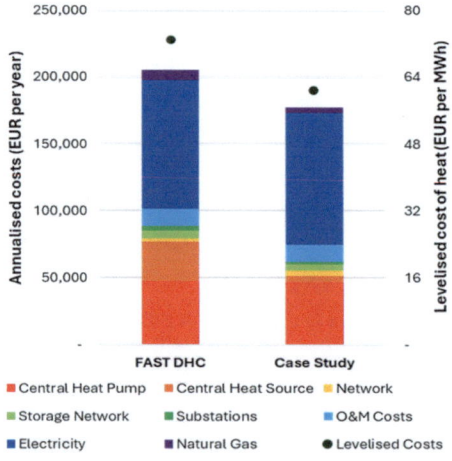

Annualised Costs	FAST DHC	Case Study
Central Heat Pump	€47,123	€46,586
Central Heat Source	€29,877	€4,685
Network	€2,487	€4,109
Storage Network	€5,305	€4,771
Substations	€3,560	€1,846
O&M Costs	€13,003	€12,891
Electricity	€96,145	€98,109
Natural Gas	€7,743	€4,048
Levelised Costs	**€73**	**€61**

Fig. 5 Breakdown of estimated costs per component for FAST DHC and the case study

than the case study), with the levelised costs of energy falling to €64 per MWh (5% higher than the case study value). Although the initial validation is promising, case study 1 represents a small, traditional 4GDH system, and further work is needed to validate the FAST DHC tool for larger networks and TSN configurations.

5 Conclusions

This paper introduced the concept and modelling approach for FAST DHC, a modular, highly-flexible web-based tool for pre-feasibility assessments of low-temperature DHC applications. The tool relies on widely used published data to provide a quick evaluation of different DHC configurations with minimal user input, with the main purpose of supporting early decision making, particularly in situations where little data is available.

Initial validation against a case study of a 4GDH system is presented, which demonstrates that the FAST DHC tool is able to replicate the results of detailed feasibility studies with an acceptable level of accuracy. However, the case study described in this paper is for a small, conventional 4GDH system, and future work is planned to test and further validate the tool with case studies of larger networks and where cooling demands are present or TSN configurations have also been investigated.

Funded by the IEA DHC Technology Collaboration Programme through Annex XIV, the FAST DHC project is an international collaboration between London South Bank University, the Austrian Institute of Technology, Aalborg University and Danfoss Climate Solutions. The project ends in July 2025 and the FAST DHC tool launch is planned for the summer of 2025.

References

1. International Energy Agency Technology Collaboration Programme on District Heating and Cooling (IEA DHC) (2024) District heating network generation definitions. Accessed online: https://www.iea-dhc.org/fileadmin/public_documents/2402_IEA_DHC_DH_generations_definitions.pdf. 27 May 2025
2. Lund H, Østergaard PA, Nielsen TB, Werner S, Thorsen JE, Gudmundsson O, Arabkoohsar A, Mathiesen BV (2021) Perspectives on fourth and fifth generation district heating. Energy 227:120520
3. Gudmundsson O, Schmidt R, Dyrelund A, Thorsen JE (2022) Economic comparison of 4GDH and 5GDH systems—Using a case study. Energy 238, Part A, 121613
4. Gjoka K, Rismanchi B, Crawford RH (2023) Fifth-generation district heating and cooling systems: A review of recent advancements and implementation barriers. Renew Sustain Energy Rev 171:112997
5. Connolly D, Drysdale D, Hansen K, Novosel T (2015) Stratego project: creating hourly profiles to model both demand and supply. Accessed online: https://heatroadmap.eu/wp-content/uploads/2018/09/STRATEGO-WP2-Background-Report-2-Hourly-Distributions-1.pdf. 29 May 2025

6. European Environment Agency (2024) Cooling degree days. Accessed online: https://www.eea.europa.eu/en/analysis/maps-and-charts/cooling-degree-days?activeTab=265e2bee-7de3-46e8-b6ee-76005f3f434f. 29 May 2025
7. European Commission (2025) EU building stock observatory. Accessed online: https://energy.ec.europa.eu/topics/energy-efficiency/energy-efficient-buildings/eu-building-stock-observatory_en#how-to-use-the-bso. 29 May 2025
8. Eurostat (2025) Average size of housing. Accessed online: https://ec.europa.eu/eurostat/cache/digpub/housing/bloc-1b.html. 29 May 2025
9. TABULA WebTool (2017) Building typologies. Accessed online: https://webtool.building-typology.eu/. 29 May 2025
10. Heat Roadmap Europe 2050 (2017) Space cooling technology in Europe. Accessed online: https://heatroadmap.eu/wp-content/uploads/2018/11/HRE4_D3.2.pdf. 29 May 2025
11. nPro Software (2025) Domestic hot water demands in districts. Accessed online: https://www.npro.energy/main/en/load-profiles/heating/domestic-hot-water. 29 May 2025
12. Eurostat (2025) Electricity price statistics. Accessed online: https://ec.europa.eu/eurostat/statistics-explained/index.php?title=Electricity_price_statistics. 29 May 2025
13. Hotmaps (2019) Hotmaps open data repositories. Accessed online: https://wiki.hotmaps.eu/en/Hotmaps-open-data-repositories. 29 May 2025
14. Sánchez-García L, Averfalk H, Möllerström E, Persson U (2023) Understanding effective width for district heating. Energy 277:127427
15. Danish Energy Agency (2025) Technology catalogues. Accessed online: https://ens.dk/en/analyses-and-statistics/technology-catalogues. 29 May 2025
16. Eurostat (2025) Purchasing power parities—database. Accessed online: https://ec.europa.eu/eurostat/web/purchasing-power-parities/database. 29 May 2025
17. Jones P, Dunham C, Fenner R, Lagoeiro H, Roszynski K, Marques C, Maidment G (2023) Energy Superhubs—the use of supermarkets as local energy centres. In: CIBSE technical symposium, Glasgow, UK, 20–21 Apr 2023
18. Hamilton J (2022) Decarbonising heat using secondary sources and Government policy—waste heat. Available online: https://ior.org.uk/public/downloads/HD2mP/Dr-Joel-Hamilton-June-22.pdf. 04 Jul 2025

Open Access This chapter is licensed under the terms of the Creative Commons Attribution 4.0 International License (http://creativecommons.org/licenses/by/4.0/), which permits use, sharing, adaptation, distribution and reproduction in any medium or format, as long as you give appropriate credit to the original author(s) and the source, provide a link to the Creative Commons license and indicate if changes were made.

The images or other third party material in this chapter are included in the chapter's Creative Commons license, unless indicated otherwise in a credit line to the material. If material is not included in the chapter's Creative Commons license and your intended use is not permitted by statutory regulation or exceeds the permitted use, you will need to obtain permission directly from the copyright holder.

Testing Dynamic Sector-Coupled Operation in a District Energy Laboratory

Oliver Gehrke, Lucas Venge Ejlsborg, and Kai Heussen

Abstract This paper discusses closed-loop dynamic system integration and interoperability testing as crucial tool in the development of sector coupled and smart district heating and cooling (DHC) systems. Sector-coupled test facilities are needed to enable such complex integrated tests across physical, automation and control layers. This paper presents three tests in detail, each representative of experiments addressing different layers of the system: (1) Characterization of physical hardware, (2) validation and calibration of the automation layer, and (3) a more complex application featuring multiple controllers across both domains. The tests, performed at the SYSLAB facility at DTU, demonstrate the feasibility of integrated testing of dynamic sector-coupled grids in operation.

Keywords Sector coupling · District heating · Laboratory testing · Smart energy systems · SGAM

1 Introduction

Sector coupling is the concept of enabling a tighter integration between the electricity and thermal sectors. From an electrical network perspective, energy exchange with the DHC sector holds tremendous flexibility potential, due to thermal inertia, cost-efficient storage and new concepts of operation [5]. Likewise, DHC utilities, already used to operating combined heat and power plants in electric power markets, look towards the deployment of large-scale heat pumps and lower-temperature operation of their networks to boost efficiency [13]. While research and field trials

O. Gehrke (✉) · L. V. Ejlsborg · K. Heussen
Department of Wind and Energy Systems, Technical University of Denmark, Roskilde, Denmark
e-mail: olge@dtu.dk

L. V. Ejlsborg
e-mail: s214276@dtu.dk

K. Heussen
e-mail: kheu@dtu.dk

have demonstrated that customer and network flexibility can offer significant savings and operational benefits [1], exploitation of these potentials requires systemic change, such as upgrading monitoring and control systems both at the network and the building level [12, 13], as well as improved design, modelling, simulation and control methods [7]. Changes of this scale require opportunities to test technologies involved in system transformation.

Similar to the testing of smart grid solutions in the electricity domain, where changing operating paradigms motivate new types of testing and test infrastructure [4, 6], the DHC domain is facing new testing requirements. Many of these are driven by an increased need for system integration and interoperability across domains and organisations. This in turn requires the ability to test complex systems-of-systems in their entirety, capturing physical phenomena at the component and network level, the performance of digital monitoring and control infrastructure, and the interaction with one or multiple controllers. As an example, a novel DH substation concept may involve new components optimized for lower temperatures, additional instrumentation and the need to interact securely with flexible consumers, energy markets or other entities.

This paper intends to showcase the needs and escalating complexity of integrated systems testing in a sector coupling context, using the example of a purpose-built sector coupling testing facility.

2 Sector Coupling Test Beds

As outlined above, sector coupling is almost never limited to the physical exchange of energy alone. For instance, a heat pump can transfer energy from an electrical grid to a district heating network, but most useful real-world applications of the unit will need operational coordination with the needs and limits of both infrastructures. The SGAM model [3], despite its origins in the electrical domain, can also be used to interpret these needs as a stack of interoperability layers for DHC and sector coupled applications [10]:

1. A component layer defining the physics of energy exchange: Pipes, valves, pumps, heat source and sinks, sensors etc.,
2. A communication layer defining the of information exchange and associated protocols: Fieldbus interfaces, cables, signals, gateways etc.,
3. An information layer defining the data representation, data models, data storage and retrieval from and to the physical system,
4. A function layer defining operational (control) logic and applications, and
5. A business layer defining business logic and regulation.

It appears logical that, in order to perform tests on integrated systems, facilities are needed to execute integrated tests across multiple of these layers and across several electrical and heating domains (e.g. grid and building domains). Currently, only a limited number of dedicated sector coupling test beds exist, including DHTF

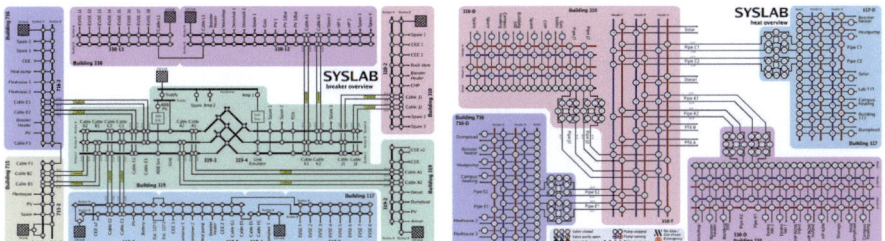

Fig. 1 Electrical (left) and thermal (right) network layout in SYSLAB. Colors represent geographically separated sites interconnected by cables and pipes

at RSE [2], CoSES at TU Munich [11], Energyville at VITO [9], and DistrictLAB at Fraunhofer IEE [8] as well as the SYSLAB facility at DTU.

2.1 The SYSLAB Sector Coupling Facility

SYSLAB is a highly observable, highly automated, sector coupled energy system research facility designed as a testbed for control and communication concepts in electrical and multi-carrier energy systems. It combines a complex physical energy system with a platform for the rapid deployment of different types of control algorithms. A 27-busbar electrical distribution grid and a district heating network with 15 header lines and 4.5 km of twin pipe form the backbone of the system, allowing interconnections between approximately 40 DER units (production, consumption, storage) in a large variety of system topologies.

All DER units as well as the grid substations are fully automated and can be monitored and controlled through a unified remote API, enabling it to emulate different kinds of real-world network topologies, such as long rural feeders, branched urban networks, islanded networks etc. This enables testing of a wide range of energy management and control concepts, and supports the automation of test execution, parameter tuning etc. The facility is spread across six different sites within an area of about 1 km^2, and the typical size of DER units in SYSLAB is in the range of 10 to 125 kVA electrical and 10 to 150 kW thermal power (Fig. 1).

3 Experimental Results and Analysis

The tests presented here represent three levels of system complexity, each building on the capabilities of the previous one:

- *Stationary characterization test*: Characterization of physical properties (heat losses) of the district heating network. This tests the physical system, i.e. the set

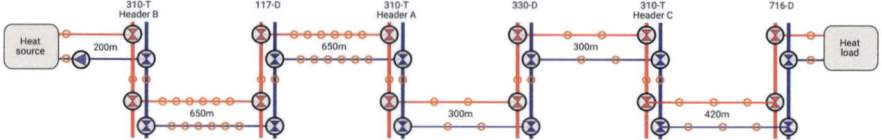

Fig. 2 System layout for the decay experiments

of hardware devices defining the basic behaviour of the test bed (corresponding to SGAM layer 1).
- *Closed-loop dynamic operation*: Simple demonstration of closed-loop control used to shape the energy transferred in the system. This test includes an automation aspect, i.e. control and automation of the test bed (including SGAM layers 1–3) rather than automation specific to a particular test case.
- *Energy community case*: Application scenario with multiple distributed controllers interacting with the physical system. This is a full system integration test, involving specific controllers-under-test as part of a larger test case (covering all SGAM layers).

3.1 Stationary Characterization

Case description This experiment belonged to a series of tests conducted to determine the basic characteristics of the physical laboratory infrastructure, and to calibrate models of the system. In this case, the coefficients of heat transfer due to (1) longitudinal diffusion and (2) conductive losses are being determined experimentally. In order to separate these two effects, two tests are needed: One to characterize the overall cooling rate of a homogenously heated system (where no significant longitudinal diffusion can occur), and another one testing the decrease of amplitude of a heat impulse moving through the system (from which the now known conductive losses can be subtracted).

Characterization of static losses To determine the cooling rate of a heated system, the entire system is homogenously heated to a preset temperature, after which all flow is stopped and the system is left to cool without interference for 3 days and 5 h. The system layout consists of a large loop of almost 5 km in length (Fig. 2).

The results are shown in Fig. 3, with an exponential decay function fitted to the data.

Characterization of Longitudinal Diffusion To determine the longitudinal diffusion rate in both directions, we measure the decrease of amplitude of a heat impulse moving through the system. A pulse of hot water is injected into the system and observed as it circulates (Fig. 4 left). To determine the decay rate of the peak, all

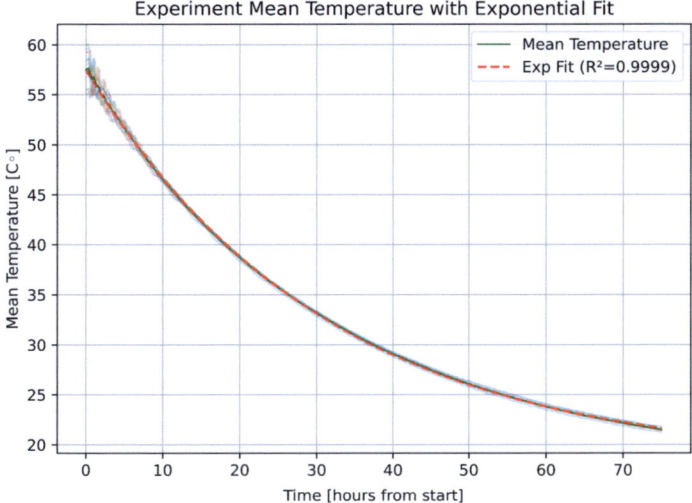

Fig. 3 Exponential function fitted to the mean of the recorded decay data from all measurement locations along the feeder

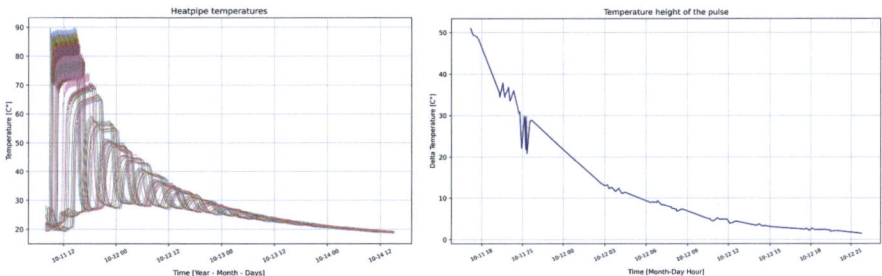

Fig. 4 Left figure: Heat wave initial data. Right figure: Temperature decay of the observed heat pulse over time (0 is the temperature equilibrium without losses)

pulse observations are aligned and the temperature differences between peak and ambient temperature are plotted as a function of time. The conductive losses are then subtracted in order to isolate the longitudinal diffusion (Fig. 4 right).

3.2 Closed-Loop Dynamic Operation

Case description This experiment was conducted as part of a series of tests designed to validate the automation layer of the laboratory. Unlike the previous case, which only required measurements to be read from the system while all interventions were

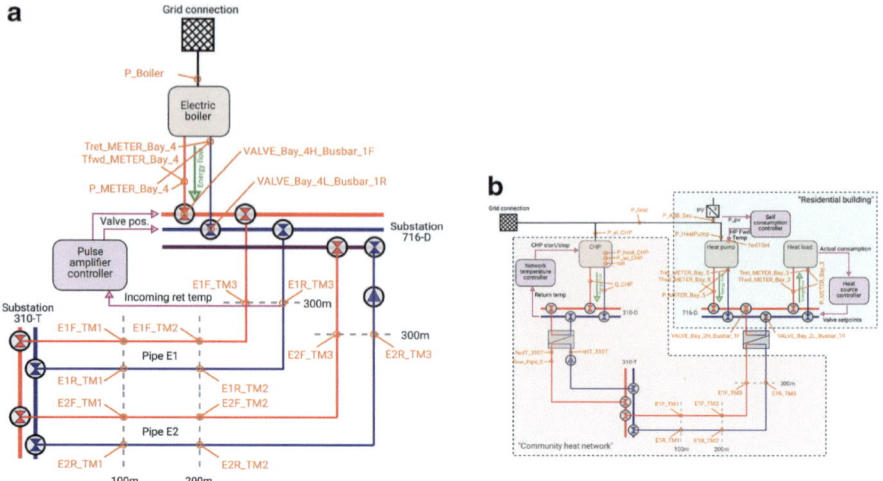

Fig. 5 **a** System configuration for the closed-loop experiment (measurement locations in orange, control signals in purple). **b** System configuration for the energy community case (measurement locations in orange, control signals in purple)

performed manually, this test requires closed-loop operation between the laboratory and a control algorithm.

The test involves creating a "pulse" of heat by injecting hot water from the storage tank of an electric boiler into a looped section of district heating network. The boiler is thermostatically controlled to maintain a constant tank temperature. There are no heat consumers connected; the only heat extraction from the system is due to thermal losses. Every time the heat pulse passes through the connection point of the storage tank, more energy is released in such a manner that the width of the pulse increases.

Experiment The system is configured as shown in Fig. 5a. In order to increase the possible width of the pulse, the network length is increased to about 800 m by employing two pipes in series, such that the flow travels twice between the two substations. In order to account for valve actuation delays, the controller detects the incoming pulse on the return line about 100 m ahead of the connection point (marked E1R_TM3) and calculates the arrival of the pulse based on the measured (and constant) pipe flow.

Results and Analysis Figure 6 contains the results from the experiment. The left subfigure shows the temperature in the forward pipe of the district heating loop at the first measurement point after the heat injection point (blue). A variable threshold calculated from a sliding window average (orange) is used to determine the "packet length" for each pass (area shaded red).

The right subfigure shows the increase of pulse length over eight full circulations. The shape of the individual pulses confirms that the controller precisely releases

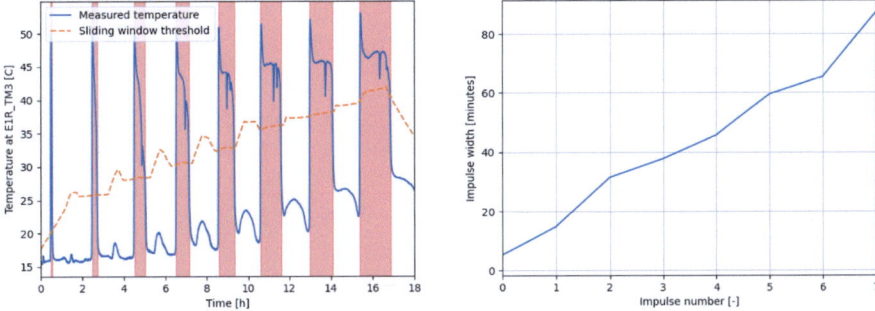

Fig. 6 Left side: Temperature time series (blue) at the first measurement point after the heat source. Right side: Detected peak length (marked with red background in the left graph) versus peak number

energy at the tail end of the plateau (growing the pulse laterally) rather than during the passage of the plateau (which would grow the pulse amplitude).

3.3 Simple Energy Community Case

Case description The last experiment is more complex control case, representative of the laboratory's intended use. It emulates a configuration which could be found in an energy community, where a residential building is connected to two heat sources: A small heat pump within the building, as well as a local heat network supplied by a CHP unit shared with other users. The residential building is also equipped with rooftop PV which is able to contribute to the supply of the heat pump behind the meter. It is assumed that electricity tariffs and taxation provide a strong incentive towards maximizing PV self-consumption. Furthermore, the PV-supplied heat pump is assumed to deliver cheaper energy than the CHP-supplied heat network.

Experiment Figure 5b shows the system configuration for the experiment. Three independent controllers govern the operation of the system:

1. A self-consumption controller (part of a building management system) adjusts the forward temperature of the heat pump according to available PV generation, in order to maximize self-consumption of PV energy.
2. A heat source controller (another part of the building management system) ensures that the heat demand of the building can be matched by a combination of the two available sources, as inexpensively as possible.
3. A network temperature controller which maintains a minimum operating temperature in the heat network by cycling the CHP.

Results and Analysis Figure 7 shows the results of the test. It can be seen that, approximately 1 h after the start of the experiment, heat pump output is no longer

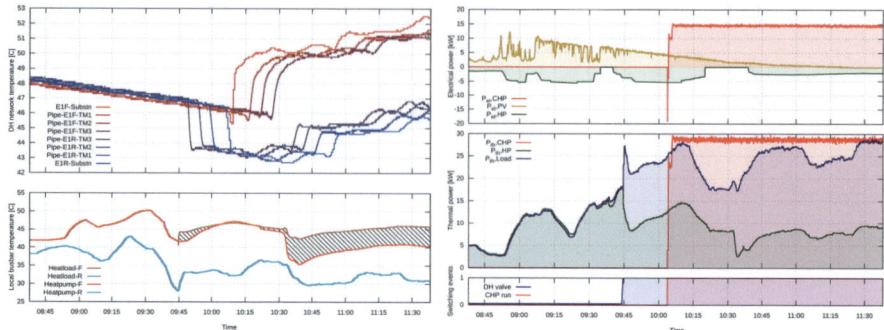

Fig. 7 Results of the energy community experiment. Top left: Heat network temperatures along the circuit. Bottom left: Heat pump and load temperatures at point of connection. Right top: Electrical production and consumption. Right center: Thermal power of heat sources and sinks. Right bottom: Switching events

sufficient to serve demand, and the heat source controller opens the heat supply from the network. As a result, a wave of cold return water propagates through the network, eventually triggering the start of the CHP unit as the network temperature drops below the minimum threshold. The output of the CHP then propagates back through the network as a wave of hot forward water.

4 Conclusion

In this paper, we have discussed the need for testing facilities to aid the development of smart DHC systems as well as sector coupling: A facility which is able to conduct complex integrated tests across the different layers present in such systems, from the physical system to communication, automation and control. Using the new electrical-thermal sector coupling test facility SYSLAB at DTU as an example, we have discussed three examples of such tests, representing three different levels of integration. The tests demonstrate the importance and feasibility of integrated testing of dynamic sector-coupled grids in operation.

References

1. Abbá I, La Bella A, Corgnati SP, Corsetti E (2024) Assessing flexibility in networked multi-energy systems: a modelling and simulation-based approach. Energy Rep 11:384–393. https://doi.org/10.1016/j.egyr.2023.11.049
2. Anderis C, Alvarado MAM, Lazzari R (2024) Implementation of an experimental facility for district heating networks flexibility assessment. In: 2024 AEIT international annual conference (AEIT), pp 1–6. https://doi.org/10.23919/AEIT63317.2024.10736713

3. CEN-CENELEC-ETSI Smart Grid Coordination Group (2014) SG-CG/M490/F overview of SG-CG methodologies. Tech Rep
4. Gehrke O, Ziras C, Heussen K, Jensen T, Bindner H (2021) Testbeds for active distribution networks: case experience from SYSLAB. In: Proceedings of the 26th international conference and Exhibition on electricity distribution (CIRED), no 6 in IET Conference Publ. https://doi.org/10.1049/icp.2021.1855
5. Golmohamadi H, Larsen KG, Jensen PG, Hasrat IR (2022) Integration of flexibility potentials of district heating systems into electricity markets: a review. Renew Sustain Energy Rev 159. https://doi.org/10.1016/j.rser.2022.112200
6. Heussen K, Steinbrink C, Abdulhadi I, Van Hoa N, Degefa M, Merino J, Jensen T, Guo H, Gehrke O, Bondy D, Babazadeh D, Andrén F, Strasser T (2019) ERIGrid holistic test description for validating cyber-physical energy systems. MDPI Energies 12(14). https://doi.org/10.3390/en12142722
7. Horak D, Hainoun A, Neugebauer G, Stoeglehner G (2022) A review of spatio-temporal urban energy system modeling for urban decarbonization strategy formulation. Renew Sustain Energy Rev 162:112426. https://doi.org/10.1016/j.rser.2022.112426
8. Kallert A, Lottis D, Shan M, Schmidt D (2021) New experimental facility for innovative district heating systems—District LAB. Energy Rep 7:62–69. https://doi.org/10.1016/j.egyr.2021.09.039
9. Koussa JA, Baeten R, Robeyn N, Salenbien R (2019). A multipurpose test rig for district heating substations Domestic hot water preparation and keep-warm function comparison. https://doi.org/10.1051/e3sconf/201911106012
10. Ma Z, Knotzer A, Billanes JD, Jørgensen BN (2020) A literature review of energy flexibility in district heating with a survey of the stakeholders' participation. Renew Sustain Energy Rev 123:109750. https://doi.org/10.1016/j.rser.2020.109750
11. Peric VS, Hamacher T, Mohapatra A, Christiange F, Zinsmeister D, Tzscheutschler P, Wagner U, Aigner C, Witzmann R (2020) CoSES laboratory for combined energy systems at TU Munich. https://doi.org/10.1109/PESGM41954.2020.9281442
12. Sporleder M, Rath M, Ragwitz M (2022) Design optimization of district heating systems: a review. Front Energy Res 10. https://doi.org/10.3389/fenrg.2022.971912
13. Østergaard DS, Smith KM, Tunzi M, Svendsen S (2022) Low-temperature operation of heating systems to enable 4th generation district heating: a review. Energy 248. https://doi.org/10.1016/j.energy.2022.123529

Open Access This chapter is licensed under the terms of the Creative Commons Attribution 4.0 International License (http://creativecommons.org/licenses/by/4.0/), which permits use, sharing, adaptation, distribution and reproduction in any medium or format, as long as you give appropriate credit to the original author(s) and the source, provide a link to the Creative Commons license and indicate if changes were made.

The images or other third party material in this chapter are included in the chapter's Creative Commons license, unless indicated otherwise in a credit line to the material. If material is not included in the chapter's Creative Commons license and your intended use is not permitted by statutory regulation or exceeds the permitted use, you will need to obtain permission directly from the copyright holder.

TFSB as Bedding Material in District Heating Pipe Construction–Scientifically Proven Long-Term Experience

Bernd Wagner, Stefan Hay, Thomas Neidhart, Florian Spirkl, Michael Ried, Louis Zrenner, Ingo Weidlich, Eugen Gabriel, and Timo Banning

Abstract Following the explanation of the increasing importance of alternative backfill materials like TFSB, TFSB and their status in terms of science and technology are depicted. Selected results of the research project FW-ZFSV 4.0 are presented: computer-aided static calculation, in situ long term loading of district heating pipes in TFSB and sand, quality assurance and resource conservation. The conclusions provide an overall assessment of the use of TFSB in district heating pipe construction.

Keywords District heating networks · TFSB · ZFSV · Design · Planning · Static calculation

1 Initial Situation

District heating competes with other heat supply solutions. Increasing technical and economic competitiveness is of major importance.

The civil engineering work for the dominant buried preinsulated bonded district heating pipes (DH pipes) can account for 55% or more of the total construction costs [1, 2]. Cost reductions, especially for excavation, backfilling and compaction offer a lever for reducing investment costs when expanding or building district heating

---

B. Wagner (✉) · S. Hay
AGFW, Frankfurt am Main, Germany
e-mail: b.wagner@agfw.org

T. Neidhart · F. Spirkl · M. Ried · L. Zrenner
OTH Regensburg, Regensburg, Germany

I. Weidlich
HafenCity Universität Hamburg, Hamburg, Germany

E. Gabriel · T. Banning
GEF Ingenieur AG, Leimen, Germany

networks. Increasing requirements of the circular economy and decreasing availability of sand are exacerbating the situation. (Keyword: Ersatzbaustoffverordnung (Substitute Building Materials Ordinance) in Germany) [3].

Alternative backfill materials are therefore becoming increasingly important [4].

2 Temporary Flowable, Self-compacting Backfill Materials

Temporary Flowable, Self-compacting Backfill Materials (TFSB) as an alternative backfill material consist of base material (soil/excavated material and/or soil building material), water and additives (hydraulic and/or layered silicate). They are free flowing after mixing and are installed without any compaction. Stiffening, solidification and hardening cause the transition to solid TFSB, which are increasingly mechanically resilient over time. Further details can be found in [5], for example.

3 TFSB in District Heating

From many years of use in sewer construction, it was and is generally known that TFSB offer potential in terms of construction technology, cost reduction, reduction of the impact of construction measures on the surrounding area and environment and conservation of resources.

Since 2009, TFSB have therefore been investigated specifically for use in district heating pipe construction in research projects. Potential savings of 3.5–16% compared to current construction technology were shown. The axial bedding mechanism was researched and clarified. Calculation approaches for displacements and forces were developed. Technical questions regarding suitability and specific requirements in the district heating sector were clarified [6, 7].

TFSB have been included in 2019 in the AGFW regulations, which is the German established code of practice for district heating and are approved for backfilling in the whole trench including the pipe area. Standards such as DIN EN 13941 District heating pipes: Design (Part 1) and Installation (Part 2) do not (yet) include TFSB.

Despite the advanced state of research and the high application potential, TFSB are mainly used for special requirements such as bottlenecks and pipe crossings. Exploration conducted in advance showed that mainly unanswered questions, a lack of solutions and uncertainties, primarily in planning and practice, hinder consideration and use as a regular alternative to conventional backfilling with sand.

4 The FW-ZFSV 4.0 Research Project

Therefore, the aim of the research project "District heating pipeline construction 4.0 with temporarily flowable self-compacting backfill materials for low and high operating temperatures" (FW-ZFSV 4.0) was to address and investigate these obstacles and provide answers to them. Specific starting points and goals were:

- Investigations into the long-term behavior of TFSB as a bedding material in district heating pipe construction, including the experimental Bypass pipe research measuring section (Bypass section) in scale 1:1 operated by AGFW in Frankfurt since 2015.
- Implementation of static calculation in planning tools.
- Answering open organizational and approval-related questions.

5 Selected Results: Static Calculation, In Situ Behavior

The project involved investigations into current issues relating to the use of the material, computer-aided static calculation, in situ long-term behavior, the use of recycled material as a base material and sustainability plus economic considerations. Selected results regarding computer-aided static calculations and in situ long-term behavior are presented in compact form below. The detailed presentation of all research results can be found in [3].

5.1 Practical Planning Tools: Computer-Aided Static Calculation

Comparative studies were carried out on simplified calculation models for initial sand bedding using several calculation programs with different solution methods.

Overall results showed that it is advantageous to calculate TFSB with beam-spring models instead of finite element models. The calculation is possible with conventional beam programs. The beam programs must be able to model the softening branch (decreasing friction) in the bedding spring that results from the underlying contact-working-resistance-line (see (Fig. 1 left). However, due to the numerous inputs in the model and for the springs, this is time-consuming, error-prone and impractical. Implementation using a specific calculation module in software explicitly designed for pipe statics is more suitable.

The implementation was therefore carried out by means of a specification sheet in software widely used in the district heating sector as bedding type TFSB (ZFSV in German) (Fig. 1 right) and will be available on the market.

Extensive comparative calculations based on the recalculation of the Bypass section confirmed the shorter length of the sliding zone and final displacements

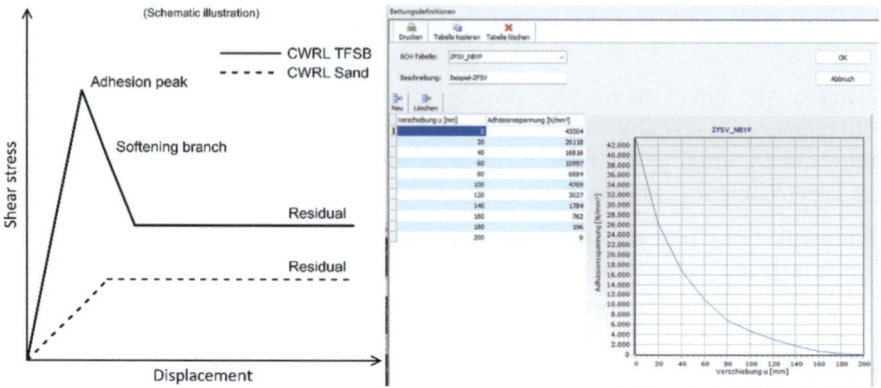

Fig. 1 CWRL: comparison of TFSB and sand (left) and implementation of bedding type TFSB in the software (right) (*Source* own illustration and screenshot)

already known from in situ displacement measurements when bedding DH pipes in TFSB compared to sand. Variation calculations showed that "softer" mixtures with a reduced adhesion peak should be aimed for in the future in the preparation of TFSB-recipes.

5.2 In Situ Long-Term Loading of DH Pipes in TFSB Bedding and Sand Bedding

Investigations were carried out on a large scale on the Bypass section in the Europaviertel district of Frankfurt am Main. The Bypass section was constructed and put into operation in 2015. Two parallel, approximately 1.6 m covered and approximately 50 m long DN 40/125 DH pipes were backfilled with TFSB respectively with sand. Under otherwise identical conditions the pipes were connected to a concrete fixed point on one side with the other end freely movable (Fig. 2). Details can be found in [7]. The Bypass section was operated by AGFW until July 2023 and then dismantled. Extensive investigations were carried out during the dismantling process.

Long-term temperature load and cyclical temperature load

The heating medium in the Bypass section has experienced temperatures of between 90 and 110 °C over several years - starting from an initial 14 °C—and has also been cooled down several times.

In the last four months of operation (Fig. 3), various cyclical temperature stresses between approximately 20–140 °C and 60–140 °C with longer holding phases at 140 °C took place. During the dismantling process, pipe sections were removed

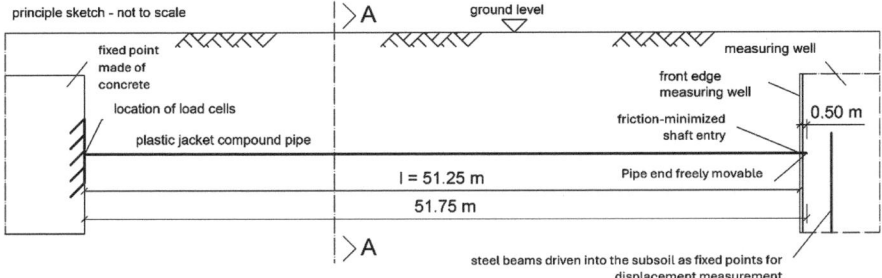

Fig. 2 Schematic longitudinal section of the Bypass section (*Source* [7], altered)

from the DH pipes for both bedding types and examined for compliance with the requirements of EN 253.

After 8 years of operation of the Bypass section results are:

- The thermal elongation impediment of the DH pipes in TFSB is greater than in sand.
- The larger thermal elongation impediment of the DH pipes bedded in the TFSB did not result in any damage to them.
- The DH pipes sections tested meet the requirements of EN 253: cell size, percentage of closed cells, water absorption at elevated temperature, foam density, compressive strength, axial shear strength.
- There are no differences due to the bedding material in the reduced bonding strength of the DH pipes caused by thermal stress.

In-situ investigation of the CWRL

In situ push-through tests of the DH pipes were carried out at several points along the Bypass section using a specifically designed, complex press construction (Fig. 4 left and middle). The tests were carried out for both bedding types under the overburden conditions of the Bypass section with an embedment length of the DH pipes of around 3 m. The aim was to determine and compare the CWRL in situ with the CWRL previously determined from Re-SIST shear tests (Fig. 4 right) in the laboratory in [7].

Figure 5 shows the CWRL for TFSB (in situ and Re-SIST) and for sand (in situ). The results in conjunction with Fig. 5 are:

- The forces and the displacements were measurable in situ; adhesion and friction can be distinguished.
- The known drop in shear stresses for TFSB takes place after the adhesion peak is exceeded.
- The in situ CWRL and the Re-SIST CWRL are similar.
- The Re-SIST test is suitable for small-scale suitability testing.
- Significantly smaller axial shear stresses occur in sand.

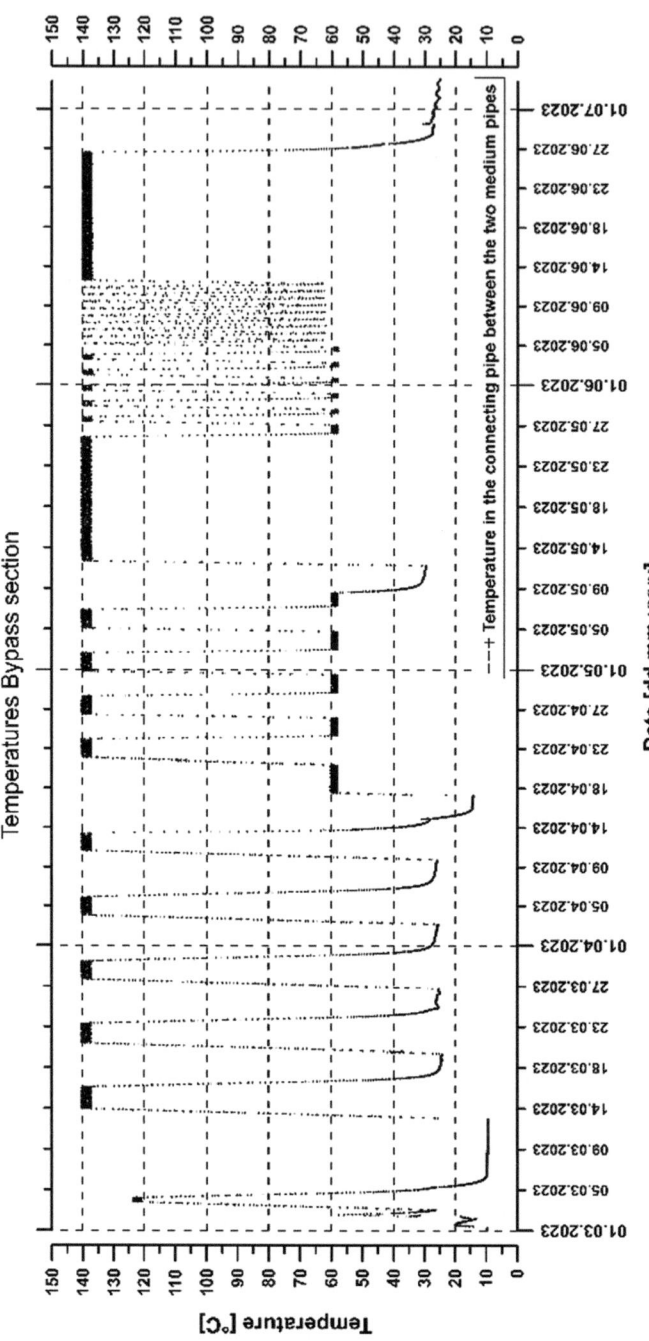

Fig. 3 Cyclical temperature load Bypass section March–June 2023 (*Source* own diagram)

Fig. 4 Press device for in situ push-through tests (left), pull and push coupling of the DH pipes with the press device (middle), Re-SIST shear test device for laboratory investigations (right) (*Source* own images)

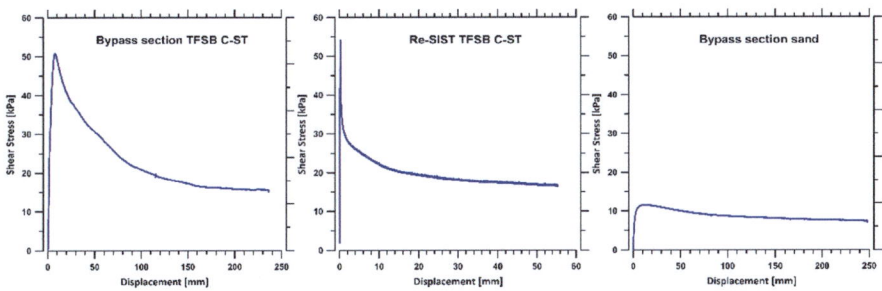

Fig. 5 CWRL TFSB in situ (left) and Re-SIST (middle) and sand in situ (right) (*Note* For display reasons, the x-axes are partly scaled differently.) (*Source* own diagram)

6 Quality Assurance and Resource Conservation

6.1 Quality Assurance

Compared to conventional pipe area backfilling with sand, the use of TFSB requires different and other quality assurance measures, which are, however, common.

When using TFSB, quality assurance must already be part of the design and planning phase. For the static design, it is necessary to decide on the characteristic CWRL value and the unconfined compressive strength regarding the re-excavation capacity. These are input values for the subsequent TFSB recipe design and suitability testing. Self-monitoring and external monitoring during the execution phase include tests to ensure that the installed material meets the specifications. The procedure and required tests are well known and standardized.

Regarding the district heating-specific quality assurance, the requirements elaborated in [7] remain mostly valid. Where necessary, they have been updated and supplementend with knowledge gained during the project in exchange with practitioners and within the research project.

It is planned to update and further anchor these measures and additional regulations on the use of TFSB in district heating in the AGFW regulations.

6.2 Resource Conservation

The greenhouse gas emissions (carbon footprint (CFP)) of sand and TFSB were compared using standard procedures and applying reasonable system boundaries.

A comparative evaluation of a real project was carried out in cooperation with a construction company. In addition, a simplified life cycle assessment was carried out.

Figure 6 left shows the results of the CFP evaluation. In Fig. 6 right the results of the simplified life cycle assessment are shown.

The assessment of the project yields that the use of TFSB as bedding material saves approximately 5% CFP (V1 versus V3). If the trench width is additionally optimized, this results in a saving of approximately 10% CFP (V1 versus V4). Additionally, one can clearly recognize in Fig. 6 right that for identical trench dimensions, the TFSB bedding only shows a higher water consumption than the sand bedding. Environmental indicators can be reduced by 30% in some cases, and by as much as 50% if the trench dimensions are optimized.

Even though the assessed project dealt with the bedding of a wastewater pipe, the authors consider the transferability of the results to other pipe systems, such as district heating pipes, to be justified.

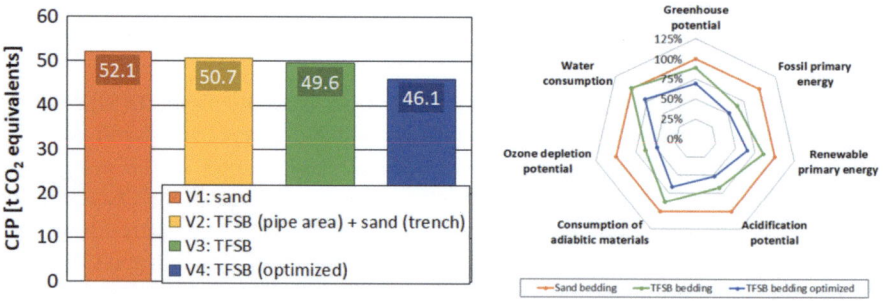

Fig. 6 Evaluated project—different construction variants: CFP for the pipe area and trench backfill (left) and simplified life cycle assessment (right) (*Source* own diagram)

7 Conclusions

In addition to the research results published to date, the results from the research project confirm that TFSB are very well suited as bedding material for district heating pipe construction and pave the way for their continued use in the future. The presented and overall results [3] of the research project yield that TFSB offer a great opportunity and a high optimization potential for expansion and construction of district heating networks. Particularly in view of the increasing demands of the circular economy, which will make the use of sand as a bedding material and the transport and disposal of excavated material even more difficult.

Acknowledgements The research project was approved and funded by the Federal Ministry for Economic Affairs and Climate Action (BMWK) under the funding code 03EN3022 via the project sponsor Jülich. The project partners are grateful for the funding, without which the implementation of the project with a 48-month term until end of July 2024 would not have been possible.

References

1. Espig F (2010) Schadensstatistik KMR 2010 des AGFW. EuroHeat&Power 41(5):32–35
2. Wagner B (2017) Ein Beitrag zur axialen Bettung von Kunststoffmantelrohren der Fernwärme in Zeitweise fließfähigen, selbstverdichtenden Verfüllbaustoffen (ZFSV). Diss. TU Bergakademie Freiberg, Freiberg
3. EnEff:Wärme: FW-ZFSV_4–0—Fernwärmeleitungsbau 4.0 mit zeitweise fließfähigen selbstverdichtenden Verfüllbaustoffen für niedrige und hohe Betriebstemperaturen—Abschlussbericht zum Verbundforschungsvorhaben. Forschung und Entwicklung | Heft 68. AGFW, Frankfurt am Main
4. Weidlich I, Dollhopf S, Hay S (2024) Alternative backfill materials for Sustainable District Heating Systems. Environmental and Climate Technologies 28(1):639–651
5. Wolfrum D (2021) Wechselwirkungsverhalten von thermisch beanspruchten Rohren und zeitweise fließfähigen, selbstverdichtenden Verfüllbaustoffen. Diss. Gottfried Wilhelm Leibniz Universität Hannover, Hannover
6. EnEff:Wärme Kostengünstiger Fernwärmetransport für den effektiven Ausbau der Kraft-Wärme-Kopplung—Bautechnische Entwicklungen von Fernwärme-Transportleitungen. Forschung und Entwicklung | Heft 32. AGFW, Frankfurt am Main
7. EnEff:Wärme: Einsatz fließfähiger Verfüllstoffe zur KMR-Verlegung. Forschung und Entwicklung | Heft 43. AGFW, Frankfurt am Main

Open Access This chapter is licensed under the terms of the Creative Commons Attribution 4.0 International License (http://creativecommons.org/licenses/by/4.0/), which permits use, sharing, adaptation, distribution and reproduction in any medium or format, as long as you give appropriate credit to the original author(s) and the source, provide a link to the Creative Commons license and indicate if changes were made.

The images or other third party material in this chapter are included in the chapter's Creative Commons license, unless indicated otherwise in a credit line to the material. If material is not included in the chapter's Creative Commons license and your intended use is not permitted by statutory regulation or exceeds the permitted use, you will need to obtain permission directly from the copyright holder.

Experimental Investigation on the Thermal Conductivity of Alternative Backfill Materials for District Heating Networks

Stefan Dollhopf and Ingo Weidlich

Abstract This study investigates the thermal conductivity of backfill materials for buried bonded pipe systems, with a focus on comparing conventional natural aggregates and recycled alternatives. Three natural and three recycled materials were tested under compacted conditions using a thermal needle probe in accordance with ASTM D5334-22a. Thermal conductivity was continuously recorded and correlated with volumetric water content. A classification methodology based on EN 13,941-1 was developed to assign thermal conductivity values to defined moisture levels (dry, medium, wet). The results show that the investigated natural materials exhibit higher thermal conductivity across all moisture states, while the investigated recycled materials demonstrate significantly lower values. Through this two-fold approach—measurement and classification—representative parameters were derived for future application in heat loss calculations. In addition, the findings could improve the understanding of the thermal interaction between soil and pipe systems, supporting the potential integration of alternative materials into sustainable infrastructure design.

Keywords Thermal soil properties · Sustainable construction methods · Recycled materials

1 Introduction

The specific heat losses [W/m] of buried bonded pipes in district heating systems depend on multiple factors, including pipe insulation, operational and soil temperatures, as well as the thermal properties of the surrounding subsoil [1–3]. While European standards define thermal conductivity coefficients for soil under dry, medium-dry, and wet conditions [4], they do not consider variations in soil composition due to the standardized use of natural sands and sand-gravel mixtures as backfill materials.

S. Dollhopf (✉) · I. Weidlich
HafenCity University Hamburg, Hamburg, Germany
e-mail: stefan.dollhopf@hcu-hamburg.de

© The Author(s) 2026
D. Vanhoudt (ed.), *Proceedings of the 19th International Symposium on District Heating and Cooling*, Lecture Notes in Networks and Systems 1700,
https://doi.org/10.1007/978-3-032-09844-3_11

However, factors such as regional regulations (e.g., Germany's substitute building materials ordinance) [5], resource scarcity [6], and the growing adoption of low-temperature networks drive interest in alternative materials [7]. Recycled materials from construction and demolition waste, such as crushed concrete, offer a promising solution aligned with resource conservation and circular economy principles. Yet, their compatibility with existing standards remains a critical concern. Therefore, it is important to improve the knowledge about these materials and evaluate the thermal properties to understand their potential impact on the thermal efficiency of district heating systems.

Within this study a two-fold approach is presented to improve the state of science for the use of alternative backfill materials in district heating networks. Thermal conductivity measurements are carried out according to ASTM D 5334–14 [8]. As the thermal conductivity of soils is highly dependent on the water saturation [9, 10] the materials are tested from a fully saturated to a dry state. Selected unbound recycled materials, including recycled concrete and brick minerals, as well as natural sands are examined under compact conditions to represent in-situ trench environments. Secondly, a classification approach is developed to determine parameters analogous to saturation-levels as defined in EN 13,941–1 [4].

2 Materials and Methods

This section presents the selection of backfill materials used in the experimental investigation, as well as the methodology of measurement, evaluation and classification of thermal conductivity.

2.1 Selected Backfill Materials

The selected materials allow for a clear comparison between conventional, natural resources and recycled alternatives. An overview of the tested materials is provided in Table 1, which includes information on grain size distributions, grain shapes, and material origins. Materials No. 1 to No. 3 represent state-of-the-art backfill materials commonly used in trench construction for district heating networks from natural resources, here mostly riversand, with round shaped grains. In contrast, materials No. 4 to No. 6 consists of processed construction waste and are generally products of a crushing process. Although these recycled materials are not currently standard for such applications in Europe due to strict normative requirements, they hold potential as sustainable alternatives in line with circular-economy principles. The material selection includes varying sources and a range of grain size distributions to ensure comprehensive representation.

The corresponding grain size curves are shown in Fig. 1. Additionally, the applicable limits for grain size distribution of bedding materials for buried bonded pipes,

Table 1 Overview of selected materials for the investigation

Number	Name	Grain-sizes	Grain-shape	Origin
1	Sand-EN489	0/3	Round	Hamburg
2	Natural-sand	0/2	Round	Rhine river
3	Sand-gravel-mix	0/10	Round	Rhine river
4	Recycling-concrete-sand	0/5	Keen-edged	Nuremberg
5	Concrete-mineral-mix	0/10	Keen-edged	Hamburg
6	Brick-mineral-mix	0/10	Keen-edged	Hamburg

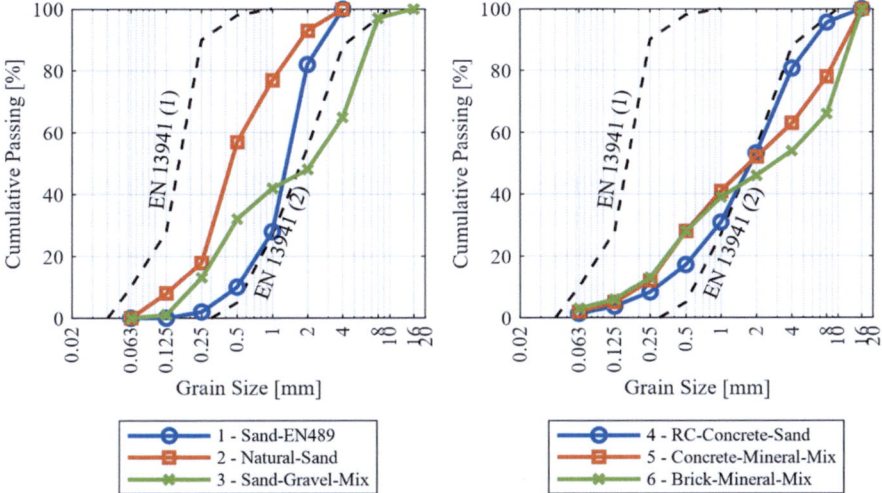

Fig. 1 Grain size distribution curves of selected materials

as defined in EN 13,941–1, were included in the graph. The materials with smaller grain fractions, Nr. 1, 2 and 4 are suitable for usage within the bedding zone. The achieved results for sand can be compared as a validation with existing measurements, like in [11–13] where thermal conductivity ranges from 1.2 to 2.2 W/(m·K) for water contents ranging from about 5 to 35 vol.%. The coarser materials Nr. 3, 5 and 6 only show minor differences within the grain distribution and offer an opportunity to evaluate the impact of the grain components on the thermal conductivity.

2.2 Procedure for the Determination of the Thermal Conductivity

The experimental investigation was conducted in accordance with ASTM D5334-22a, which outlines the standard test method for determining the thermal conductivity

Fig. 2 Measurement of thermal conductivity and sample weight with VARIOS Device by METER

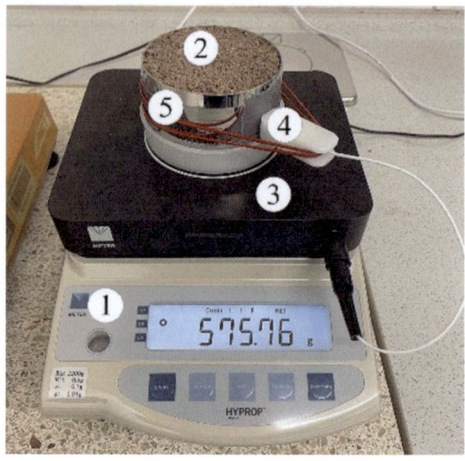

of soils and soft rocks using the thermal needle probe technique. Measurements were performed using the VARIOS device by METER Group, which combines thermal conductivity measurements with continuous weight tracking throughout the drying process. The device setup is illustrated in Fig. 2. The setup consists of the balance (1) for tracking the weight of the sample (2), the VARIOS box (3) with the thermal penetration needle (4) and the soil-sample ring (5).

All tests were carried out in the laboratory of HafenCity University Hamburg under controlled ambient conditions of 20 ± 2 °C. For each material, three specimens were prepared and tested. Initially, the samples were placed into cylindrical sample rings and compacted to reach a dense state. The specimens were then saturated with tap water via capillary suction. Once full saturation was achieved, the thermal needle was inserted into the specimen and the measurement sequence was initiated.

The VARIOS device continuously recorded both thermal conductivity and sample weight at hourly intervals throughout the drying phase, which lasted between 15 and 25 days per specimen depending on the material. Following the drying cycle, each sample was oven-dried at 105 °C for 24 h to determine its dry mass, allowing the determination of the initial moisture and the dry density. To ensure accurate weighing, the weighing system was verified based on the manufacturer's instructions to calibrate the balance. Additionally, the measurement setup was validated using a glycerine reference probe to confirm the accuracy of the thermal conductivity readings.

Complementary to the thermal measurements, grain size distribution tests were conducted for all materials in accordance with DIN EN ISO 17892–4, ensuring a comprehensive characterization of the samples.

2.3 Evaluation of Measurements

Based on the measurement data provided by the METER software, thermal conductivity curves were generated as a function of volumetric water content for each tested material. For each of the three specimens per material, individual curves were created, from which a median curve was derived to represent the characteristic thermal conductivity behavior of the material.

Due to material-specific differences, particularly in grain size distribution, pores and the presence of physically bound water in recycled materials, a direct comparison of thermal conductivity values at fixed volumetric water contents was not feasible. As a result, a classification system was developed, based on moisture levels, in alignment with the classification approach of EN 13,941–1.

The classification and evaluation were carried out using MATLAB. Initially, the raw data obtained from the VARIOS device, measured at hourly intervals, were imported. To enable consistent comparison, the thermal conductivity values were linearly interpolated to predefined steps of 0.5 vol% within the measured water content range. The typical range spanned from approximately 1–3 vol. % (dry state) to 25–40 vol. % (fully saturated state), depending on the material.

This saturation range was divided into three classes, dry state, medium wet and wet state. The mean thermal conductivity within each of these sections was then calculated to characterize the material's behavior at representative moisture levels.

3 Results and Discussion

This section presents the results of the thermal conductivity measurements performed on the selected backfill materials. Table 2 summarizes the mean dry densities, along with the minimum and maximum values for volumetric water content and thermal conductivity derived from the calculated median curves. Figure 3 illustrates the resulting thermal conductivity curves and Table 3 provides the classified thermal conductivity values at three moisture levels, as outlined in Sect. 2.3.

The dry density results show a close alignment between the two sand materials (materials 1 and 2), with only minor deviations. Both achieve typical values indicative of densely packed sand layers. Material 3, a sand-gravel mix, exhibits a higher dry density, which is attributed to its broader grain size distribution that facilitates denser compaction. This finding confirms the compactability advantage of well-graded granular materials. In general, materials derived from natural resources (Materials 1–3) exhibit higher dry densities compared to recycled materials (Materials 4–6). This can be explained by the lower intrinsic grain density and the presence of internal porosity in recycled aggregates.

The volumetric water content of the sands ranges from approximately 0 vol% to 31–35 vol%, due to the improved compaction of the sand-gravel mix, its maximum

Table 2 Overview of mean dry density as well as minimum and maximum values of volumetric water content and thermal conductivity for all investigated materials

Nr	Name	Dry density [g/cm³]	Water content [vol. %]		Thermal conductivity [W/m*K]	
			Min	Max	Min	Max
1	Sand-EN489	1.71	0.5	31.5	0.4	2.3
2	Natural-sand	1.73	0.0	35.0	0.3	2.3
3	Sand-gravel-mix	1.97	0.0	23.5	0.3	2.5
4	Recycling-concrete-sand	1.54	3.5	31.5	0.2	1.4
5	Concrete-mineral-mix	1.65	5.0	32.5	0.3	1.8
6	Brick-mineral-mix	1.40	3.0	38	0.2	1.2

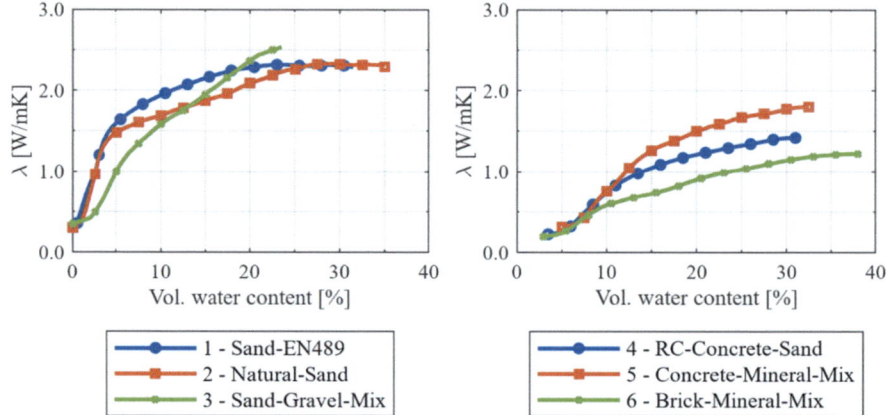

Fig. 3 Thermal conductivity as a function of volumetric water content for all investigated materials (median values derived from three specimens each) at dense state

volumetric water content is lower. An interesting observation is that none of the recycled materials reached a volumetric water content of 0 vol% under ambient drying conditions, with minimum values remaining above 3 vol.%. Complete desaturation was only achieved through oven drying at 105 °C, suggesting the presence of physically or chemically bound water within the grain structures. The maximum water contents of the recycled materials, however, are comparable to those of the natural materials, implying a broadly similar particle-size distribution. In particular, the elevated water content of the brick-mineral mix (Material 6) can be attributed to the high porosity of the constituent grains, which are capable of absorbing and retaining water. This characteristic explains the material's susceptibility to frost damage and restricts its use to non-frost-prone environments.

The thermal conductivity curves in Fig. 3 reveal that natural sands exhibit a sharp drop in thermal conductivity at around 5 vol% water content. This phenomenon is

Table 3 Thermal conductivity classification of backfill materials according to moisture levels (dry, moist, and wet) in reference to EN 13,941-1

	EN 13,941-1 (Reference)	1–Sand-EN489	2–Natural-sand	3–Sand-gravel-mix	4–RC-Concrete-sand	5–Concrete-mineral-Mix	6–Brick-mineral-mix
Dry	1.0	**1.5**	1.3	0.8	0.5	0.7	**0.5**
Medium wet	1.6	**2.2**	2.0	1.7	1.1	1.4	**0.9**
Wet	2.0	**2.3**	**2.3**	**2.3**	1.4	1.7	**1.1**

likely due to the loss of continuous water films between grains, which significantly reduces heat transfer. This effect is less pronounced in the sand-gravel mix (Material 3), where a more diverse grain size distribution provides more persistent grain-to-grain contacts and capillary water retention. The minimum thermal conductivity for the natural materials is around 0.3 W/(m·K), with maximum values between 2.3 and 2.5 W/(m·K).

In contrast, all recycled materials exhibit significantly lower thermal conductivities than their natural counterparts, with values ranging from approximately 0.2 W/(m·K) to a maximum of 1.2–1.8 W/(m·K). Furthermore, the curves of the recycled materials show a more linear increase in thermal conductivity with rising water content.

Overall, the measured values are consistent with previous studies and align with the physical understanding that internal porosity in recycled aggregates—such as crushed concrete or brick—tends to reduce thermal conductivity due to disrupted heat transfer pathways within the grain matrix.

The classified thermal conductivities of the investigated backfill materials at defined moisture levels are presented in Table 3. In comparison with the reference values provided in EN 13,941–1, the results confirm that the standard values are broadly applicable to a range of soil and granular materials. For conventional backfill materials (materials 1 to 3), the determined thermal conductivities mostly exceed the EN 13,941–1 reference values. The highest deviations can be found in the medium-wet state of material 1, where the reference value of EN 13,941–1 is 1.6 W/(m·K) and the determined values range from 1.7 to 2.2 W/(m·K).

In contrast, the recycled materials (materials 4–6) generally exhibit lower thermal conductivity. For instance, all recycled materials show values in the wet state (1.1 to 1.7 W/(m·K) that are significantly below the reference value of EN 13,941–1 (2.0 W/(m·K)).

4 Conclusion

This investigation contributes to improving the current understanding of the impact on the thermal behavior of a greater variation of materials used as backfill materials for buried bonded pipes in district heating networks.

The thermal conductivity of each material was determined as a function of volumetric water content using a standardized procedure presented in ASTM-D 5334-22a. The results obtained for the natural materials, particularly for sand and sand-gravel mixtures, are consistent with findings from previous studies, confirming their predictable thermal performance across a range of moisture levels. The recycled materials of this study exhibited lower thermal conductivities throughout the full range of moisture conditions. Among them, the brick-mineral mix showed the lowest thermal performance, followed by the RC-concrete sand and the concrete-mineral mix. These differences can be attributed to the higher internal porosity and lower grain density of recycled aggregates, which reduce their capacity for thermal conduction, particularly in dry conditions.

The measurement data was further processed using a classification procedure to derive thermal conductivity values for defined moisture states (dry, moist, and wet), following the structure of EN 13,941–1 for the use in the heat loss determination. The resulting values align well with the standard's guidance and offer a more nuanced parameter basis for the calculation of heat losses in district heating networks. Moreover, the classification reflects the influence of both material structure and origin on the thermal behavior of the pipeline and soil interaction. These findings suggest that, under certain boundary conditions, the use of recycled materials could lead to reduced heat losses. However, the application of such materials must also meet all environmental and technical requirements, such as groundwater conditions, load-bearing capacity and frost resistance. To support broader implementation, further research is recommended. This should include the investigation of additional thermal parameters such as specific heat capacity, as well as studies on the effect of these materials on the specific heat losses of district heating networks.

Acknowledgements This work is a result of the research project UrbanTurn (03EN3029F) which is nationally funded under the EnEff:Wärme Program by the Federal Ministry for Economic Affairs and Climate Action of Germany.

References

1. Schuchardt GK, Weidlich I (2017) Sensitivity analysis of the conception of small scale district heating networks on the thermal conductivity of the surrounding soil. Energy Procedia. https://doi.org/10.1016/j.egypro.2017.09.028
2. Dalla Rosa A, Li H, Svendsen S (2011) Method for optimal design of pipes for low-energy district heating, with focus on heat losses. Energy. https://doi.org/10.1016/j.energy.2011.01.024
3. Perpar M, Rek Z, Bajric S, Zun I (2012) Soil thermal conductivity prediction for district heating pre-insulated pipeline in operation. Energy. https://doi.org/10.1016/j.energy.2012.06.037
4. CEN: EN 13941–1:2019+A1:2021—District heating pipes - Design and installation of thermal insulated bonded single and twin pipe systems for directly buried hot water networks—Part 1: Design
5. Federal Ministry for the Environment, Nature Conservation, Nuclear Safety and Consumer Protection (BMUV): Waste Management in Germany 2023
6. Peduzzi P (2014) Sand, rarer than one thinks. Environ Dev. https://doi.org/10.1016/j.envdev.2014.04.001
7. Weidlich I, Grajcar M (2017) Expected potential of bound and recycled backfill material in low temperature district heating networks. Energy Procedia. https://doi.org/10.1016/j.egypro.2017.09.035
8. D18 Committee: Test method for determination of thermal conductivity of soil and rock by thermal needle probe procedure. ASTM International, West Conshohocken, PA
9. Bertermann D, Müller J, Freitag S, Schwarz H (2018) Comparison between measured and calculated thermal conductivities within different grain size classes and their related depth ranges. Soil Syst. 2:50
10. Miles SK (1949) Thermal properties of soils. Retrieved from the University Digital Conservancy. https://hdl.handle.net/11299/124271

11. Bristow KL (1998) Measurement of thermal properties and water content of unsaturated sandy soil using dual-probe heat-pulse probes. Agric For Meteorol. https://doi.org/10.1016/S0168-1923(97)00065-8
12. Verschaffel-Drefke C, Schedel M, Balzer C, Hinrichsen V, Sass I (2021) Heat dissipation in variable underground power cable beddings: Experiences from a real scale field experiment. Energies. https://doi.org/10.3390/en14217189
13. Henoegl O, Leonhardt R, Riedler E, Honarmand H (2009) Thermal propagation around heat supply pipes—determining thermal conductivity of soil specimens. In: Hamza M, Shahien M, El-Mossallamy Y (eds) Proceedings of the 17th international conference on soil mechanics and geotechnical engineering

Open Access This chapter is licensed under the terms of the Creative Commons Attribution 4.0 International License (http://creativecommons.org/licenses/by/4.0/), which permits use, sharing, adaptation, distribution and reproduction in any medium or format, as long as you give appropriate credit to the original author(s) and the source, provide a link to the Creative Commons license and indicate if changes were made.

The images or other third party material in this chapter are included in the chapter's Creative Commons license, unless indicated otherwise in a credit line to the material. If material is not included in the chapter's Creative Commons license and your intended use is not permitted by statutory regulation or exceeds the permitted use, you will need to obtain permission directly from the copyright holder.

Achieving Efficient District Heating Targets in a Croatian Network: Heat Source Mapping and Techno-Economic Scenarios Analysis

Daniele Anania, Josip Miškić, Tomislav Pukšec, and Marco Cozzini

Abstract This study presents a replicable methodological framework for supporting the decarbonization of district heating (DH) systems, addressing both technical and economic aspects. The approach integrates spatial mapping of renewable energy sources (RES) and waste heat (WH), pre-screening of individual technologies through parametric analyses, and hourly aggregated simulation of decarbonization scenarios. The pre-screening step helps identify promising technologies early on and reduces the number of scenario simulations needed. The aggregated model accounts for temperature-dependent heat losses and dispatch priorities, and is calibrated using real operational data to ensure accurate performance representation. The methodology combines source mapping, technology economic evaluation, and scenario simulation in a structured workflow designed to support early-stage planning and local decision-making in line with the desired decarbonization targets. The methodology is applied to the DH network of Vukovar in Croatia, which is currently reliant on natural gas and features a 3% solar thermal contribution. In 2023, the total heat production of the network amounted to 12.7 GWh. RES and WH options—such as river water, air-source and shallow geothermal heat pumps, supermarket waste heat, and solar thermal—are assessed based on local availability and expected performance. Among the scenarios investigated, the one combining an extension of the solar field size up to 3000 m^2, a 1.25 MW river-source heat pump, and a 5 MWh thermal storage unit emerged as the most cost-effective solution capable of achieving the target of a 50% RES + WH share—aligned with the 2035 European definition of Efficient District Heating and Cooling. This scenario also proves to be economically

---

D. Anania (✉) · M. Cozzini
Institute for Renewable Energy, EURAC Research, Bolzano, Italy
e-mail: daniele.anania@eurac.edu

D. Anania
Energy Department, Politecnico Di Torino, Torino, Italy

J. Miškić · T. Pukšec
Faculty of Mechanical Engineering and Naval Architecture, Department of Energy, The University of Zagreb, Power Engineering and Environmental Engineering, Zagreb, Croatia

© The Authors(s) 2026
D. Vanhoudt (ed.), *Proceedings of the 19th International Symposium on District Heating and Cooling*, Lecture Notes in Networks and Systems 1700,
https://doi.org/10.1007/978-3-032-09844-3_12

competitive, with a Levelized Cost of Heat (LCOH) below the current district heating tariff.

Keywords Efficient district heating · Solar thermal · Heat pumps · Waste heat · Techno-economic scenario analysis

1 Introduction

The decarbonization of the heating and cooling (H&C) sector is a key objective of the European Union (EU). The revised Energy Efficiency Directive (EED) requires Member States to promote high-efficiency, renewable-based district heating and cooling (DHC), and from 2030 only systems meeting the definition of efficient DHC will be eligible for public support. This transformation requires strategic planning, spatial mapping of renewable energy sources (RES) and waste heat (WH), and scenario-based assessments aligned with local infrastructure and policy targets. The LIFE HeatMineDH project supports this transition by helping stakeholders to operationalize EED requirements in four European countries (Croatia, Germany, Italy, Poland).

Decarbonizing district heating (DH) systems requires a comprehensive understanding of local renewable energy sources (RES) and waste heat (WH) potentials, making the mapping phase a crucial initial step. This phase has been extensively addressed in several studies [1–5], which provide assessments of local heat availability. As outlined in [6], the integration of multiple heat supply units into multi-source DH systems has been investigated in literature, focusing on renewable energy integration, techno-economic analyses, system configurations, and optimization of control strategies. Moreover, the topic of decarbonization has been explored through long-term scenario analyses that assess system dynamics and sustainability until 2050 [7]. However, these approaches focus primarily on system operation and do not offer a unified, replicable workflow that links the phases of source mapping, technology evaluation, and scenario simulation.

The present work aims to address this gap by proposing an integrated and replicable methodology that encompasses all these stages, providing a general framework for the technical and economic assessment of DH decarbonization pathways. The methodology is applied to the City of Vukovar (Croatia), covering the entire process from heat source mapping to scenario simulation and cost evaluation. Central to the approach is an aggregated hourly energy model that includes network temperature effects—critical for technologies such as solar thermal and heat pumps—and employs a heat dispatch strategy based on source priority. The methodological framework also introduces a pre-screening step conducted after source mapping and before the scenario analysis. Here, each technology undergoes parametric techno-economic assessment to identify the most promising options based on decarbonization targets and economic competitivity. This step reduces the scenario space to be further investigated. The proposed methodology aims to support early-stage planning and aids in identifying cost-effective pathways to achieve the renewable and waste heat share required for efficient DHC.

1.1 Vukovar Case Study

The city of Vukovar, located in eastern Croatia along the Danube River, is partially supplied by a district heating system that covers a portion of the city's heat demand. In 2023, the system produced a total of 12.7 GWh of heat, of which about 12.4 GWh (97%) was generated from natural gas, using three boilers with a combined installed capacity of 18.2 MW. The remaining 0.3 GWh was provided by a solar thermal field consisting of flat-plate collectors, with a total aperture area of approximately 850 m^2 and an installed capacity of 0.52 MW. Heat is distributed through a 5.5 km-long network operating with variable supply temperatures, reaching up to 95 °C in winter and dropping to around 65 °C in summer. The average temperature difference between supply and return is approximately 20 K. In 2023, the total heat demand of connected users amounted to 10.9 GWh, indicating network thermal losses of around 14%.

These figures highlight that the current district heating system is far from being compliant with the definition of efficient DHC reported in the EED, even under the current requirements, which are set to become more stringent in the coming years. The focus of this study is to identify a cost-effective combination of heat sources that can achieve a 50% share of renewable and waste heat by 2035, in line with future efficiency targets.

2 Methodology

The proposed methodological framework is structured in three steps: (i) mapping available local heat sources, (ii) conducting a techno-economic pre-screening to focus on the most promising heat source configurations, and (iii) performing multi-source scenario simulations with an aggregated hourly model. The scenario analysis aims to assess system performance, decarbonization targets, and economic viability. In this study, the pre-screening is carried out in two stages: first, by varying solar collector area and TES capacity to identify promising solar setups, and then by adding heat pumps to assess combined renewable share targets and cost-effectiveness. This approach efficiently directs computational resources toward the most viable solutions, avoiding simulations of scenarios lacking competitiveness.

2.1 Heat Source Mapping

QGIS was used to map renewable and waste heat sources near the existing DH network. The focus was on unbuilt areas suitable for solar thermal collectors and shallow geothermal systems with HPs, low-temperature WH (e.g., supermarkets) exploitable via HPs, and water bodies for water-source HPs. The WH mapping followed the procedure of Ref. [8], estimating the annual recoverable heat of supermarkets based on their floor area. Potential shallow geothermal extraction was calculated assuming double U-tube boreholes with a 6 m spacing and assuming 50 W/m of extraction rate and a geothermal gradient of 0.05 °C/m, in line with the geothermal conditions observed in Croatia [9]. The geothermal gradient is also confirmed by a site-specific investigation in Vukovar. Assuming a borehole depth of 150 m (typical value for closed-loop applications), a source temperature of 15 °C is expected. For solar thermal, the potential was estimated using a quasi-dynamic model calibrated on monitoring data from the existing Vukovar solar field, details are in Sect. 2.3. The Danube River has a minimum daily flow of 1300 m^3/s (2023 data [10]). Assuming a conservative temperature drop of 3 K and a 1% flow usage, a power of the order of 100 MW could be obtained. This rough estimation simply shows that the river can support large-scale heat pump operation with no capacity constraints for the network considered in this study.

2.2 Aggregated District Heating Model

The DH system is modelled using an aggregated, hourly energy balance approach. The main inputs to the model are the hourly profiles of aggregated user heat demand Q_{usr}, supply and return temperatures of the DH network, and ambient air temperature. At each timestep θ, the total heat demand—including user consumption and network losses—is covered by a combination of heat sources $\mathcal{S}$ and thermal storages $\mathcal{T}$, following a hierarchical dispatch strategy. The energy balance is formulated as:

$$\sum_{i \in \mathcal{S}} Q_i(\theta) + \sum_{j \in \mathcal{T}} Q_j(\theta) = Q_{usr}(\theta) + Q_{loss}(\theta) \quad (1)$$

where Q_i is the heat production of the source $i \in \mathcal{S}$, and Q_j the contribution of storage $j \in \mathcal{T}$, positive when discharging (i.e., acting as an additional heat source) and negative when charging (i.e., acting as an additional heat load). Network heat losses are computed as:

$$Q_{loss}(\theta) = UA_{ntw} \cdot (\overline{T}_{ntw}(\theta) - T_g(\theta)) \quad (2)$$

with $\overline{T}_{ntw}$ the average network temperature and T_g the surrounding ground temperature at the pipe depth, estimated from ambient temperature and depth using the correlation reported in [11].

Sources are grouped into primary $\mathcal{P}$ and auxiliary $\mathcal{A}$ units. The dispatch follows a hierarchical structure, prioritizing primary sources first, then thermal storages, and finally auxiliary sources. Within each group, units are activated based on internal priority. When the heat available from primary sources exceeds the demand, the surplus is used to charge thermal storages, limited by storage capacity and maximum charging power. Conversely, when primary sources cannot meet the demand, storages are discharged to cover the gap. If this is still insufficient, auxiliary units are progressively activated. The actual output of each unit depends on the requested thermal power, the unit's maximum available capacity, and minimum modulation thresholds.

This control strategy is designed to maximize the share of heat supplied by primary sources. In line with the decarbonization targets defined for the Vukovar DH system, renewable technologies such as solar thermal, heat pumps, and waste heat are classified as primary sources, whereas natural gas boilers are treated as auxiliary units and dispatched only when necessary.

2.3 Heat Source Modelling

The thermal power available from the solar thermal field is computed as:

$$Q_{solar}(\theta) = \left(G_{tilted}(\theta) \cdot \eta_c(\theta) - k \cdot \Delta T_{w,a}(\theta)\right) \cdot A_{tot} \tag{3}$$

where G_{tilted} is the solar irradiance incident on the tilted surface, which depends on geographic location, collector orientation, and timestep θ. It is estimated according to the methodology described in [12], using climatic data from [13]. The efficiency of solar collectors η_c is calculated according to the European Standard EN 12975 as follows:

$$\eta_c(\theta) = \eta_0 - a_1 \frac{T_m(\theta) - T_a(\theta)}{G_{tilted}(\theta)} - a_2 \frac{(T_m(\theta) - T_a(\theta))^2}{G_{tilted}(\theta)} \tag{4}$$

The efficiency parameters η_0, a_1, and a_2 are taken from the datasheets of the collectors installed in the current solar field and are equal to 0.83, 3.29 W/(m²·K), and 0.04 W/(m²·K²), respectively. The hourly ambient temperature profile T_a is provided as an input to the model, while the medium fluid temperature in the collectors T_m is computed as:

$$T_m(\theta) = \frac{T_{out}(\theta) - T_{in}(\theta)}{2} \tag{5}$$

where T_{out} and T_{in} correspond to the supply and return temperatures of the district heating, respectively. In Eq. (3), $\Delta T_{w,a}$ represents the temperature difference between the water in the collector and the ambient air, and A_{tot} is the total aperture area of the solar field. The term k is a calibration factor introduced to account for thermal losses in the piping network of the solar field. This factor was calibrated using monitoring data from the existing Vukovar's solar thermal plant, yielding a value of 3.18 W/($m^2 \cdot K$).

In the model, the heat production available from a HP is defined as the product of its nominal heating capacity and a Boolean availability profile, which is zero during downtime such as maintenance. HP operates above a minimum modulation threshold of 20%. Electricity consumption is estimated using the COP, calculated with the Eq. (4) proposed in [14], using parameters in line with the case studies presented in [15] (maximum COP deviation of approx. 10%).

$$\text{COP} = \eta_r \cdot \left(1 - \eta_m + \eta_m \cdot \frac{T_{out,co} + \Delta T_{hex,co}}{T_{out,co} - T_{out,ev} + \Delta T_{hex,co} + \Delta T_{hex,ev}}\right) \quad (6)$$

The term $\eta_m = 0.55$ can be loosely interpreted as the mechanical efficiency of the compressor, while $\eta_r = 0.90$ denotes the system integration efficiency, which accounts for real-world installation effects. Temperature differences at heat exchangers ΔT_{hex} are 2.5 K at the condenser, and at the evaporator 2.5 K for water sources and 7.5 K for air sources. The condenser outlet temperature $T_{out,co}$ equals the district heating supply temperature, while the evaporator outlet temperature $T_{out,ev}$ is based on the source temperature (ambient air, river water, shallow geothermal, or low-temperature waste heat) reduced by a temperature difference of 3 K for water sources, 5 K for geothermal sources, and 7 K for air sources. For the modelling of the geothermal source the ground temperature drift due to the heat extraction was neglected. This slightly overestimates the performance of the geothermal HP affecting the COP by a few percentage points which are considered acceptable for this application.

The annual share of each heat source is computed as the ratio between the total thermal energy it provides and the total heat production from all sources.

2.4 Economic Analysis

The parameters used for the economic analysis of the new heat sources are summarized in Table 1, including specific investment costs and operation & maintenance (O&M) expenses. For comparison purposes, gas boiler costs are reported as well.

Cost data for the HPs and gas boilers are taken from the DEA Technology Data Catalogue (2023), adjusted by a factor of 0.8 to reflect the difference in equipment price levels between Denmark and Croatia, based on the European equipment price indices [16]. Solar thermal field costs correspond to investments already sustained by the Vukovar DH operator, adjusted for inflation (a 30% increase from 2019 to

Table 1 Economic parameters for heat source technologies. Investment cost includes equipment, installation, and, for heat pumps, electrical grid connection

Water source HP	Investment cost		O&M fixed	O&M variable
	1.21	M€/MW	2126.7 €/MW	2.86 €/MWh
Air source HP	1.40	M€/MW	2126.7 €/MW	2.86 €/MWh
Solar thermal	$563.6 \cdot A^{-0.075}$	€/m2	0.04 €/MW	0.27 €/MWh
TES	1000	€/m3	/	/
Shallow geothermal	58.06	€/m	3%	/
Gas boiler	0.26	M€/MW	2658.4 €/MW	2.23 €/MWh

(*) Water source HP includes both river and geothermal heat pumps; additional costs for geothermal borehole installation are reported separately under "Shallow geothermal." O&M costs for thermal energy storage (TES) are included with the solar thermal costs.

2025 based on Eurostat data for Croatia). Thermal energy storage (TES) costs are based on Italian case-studies, whit 1000 €/m^3 applying to volumes up to 1000 m^3, corresponding to approx. 23 MWh capacity at a $\Delta T = 20$ K. Shallow geothermal data refer to double U-tube ground heat exchangers [17]. Investment costs for new dedicated DH connections are estimated based on actual data collected from existing DH developments in Italy. Pipe diameters are calculated according to transmission capacity, assuming $\Delta T = 20$ K and typical flow velocities (0.75–3 m/s).

In Croatia, the average energy prices in 2024 were 216.2 €/MWh for electricity and 51.9 €/MWh for natural gas [18]. The Vukovar DH heat selling tariff was 76.9 €/MWh for residential users and 87 €/MWh for non-residential users [19], with an average of about 82 €/MWh.

The LCOH is calculated for individual new heat sources as the ratio of total annualized costs (including investment, O&M, energy costs, and any DH connection costs) to the annual heat output provided by that source. For scenarios involving multiple new heat sources, an equivalent LCOH is computed as the total annualized costs from all new sources divided by the total heat produced by those sources combined. Annualization of investment costs is performed using a WACC of 6% and a depreciation period of 25 years. The aim of the study is to identify decarbonization scenarios capable of achieving a 50% share of RES + WH while ensuring economic competitiveness. Therefore, the LCOH of the proposed scenarios must not exceed the average heat tariff of 82 €/MWh.

3 Scenario Analysis and Results

For the city of Vukovar and the considered solar thermal panels, simulations yield a peak production of 0.65 kW/m^2 and an annual energy yield of 390 kWh/m^2. As regards instead the HP performance, the estimated seasonal COP is 2.6 for an air-source HP, 2.9 for a river-source HP using the Danube as the source, and approx. 3.0 for a geothermal HP with a source temperature of 15 °C.

The mapping of renewable heat sources revealed some promising opportunities in the proximity of the district heating network, Fig. 1.

A river-source heat pump could be installed along the Danube River, near the local drinkable water production plant. This intervention would require approx. 830 m of additional network connection and could also allow for the connection of new users, including a sport centre with potentially significant heat demand. North of the existing heat plant, an unbuilt area of 12,800 m^2 was identified. For the installation of a solar thermal field, a maximum land use of 40% is assumed to account for collector spacing and technical access, resulting in about 5,000 m^2 of maximum collector aperture area. For shallow (150 m) geothermal systems, assuming that 80% of the free area can be exploited, the maximum extractable heat at 15 °C is estimated at 2 MW (based on the parameters reported in the methodology section), corresponding to a geothermal HP capacity of 3 MW (COP = 3). In the southwestern area, a supermarket was identified as a potential source of low-temperature waste heat, with

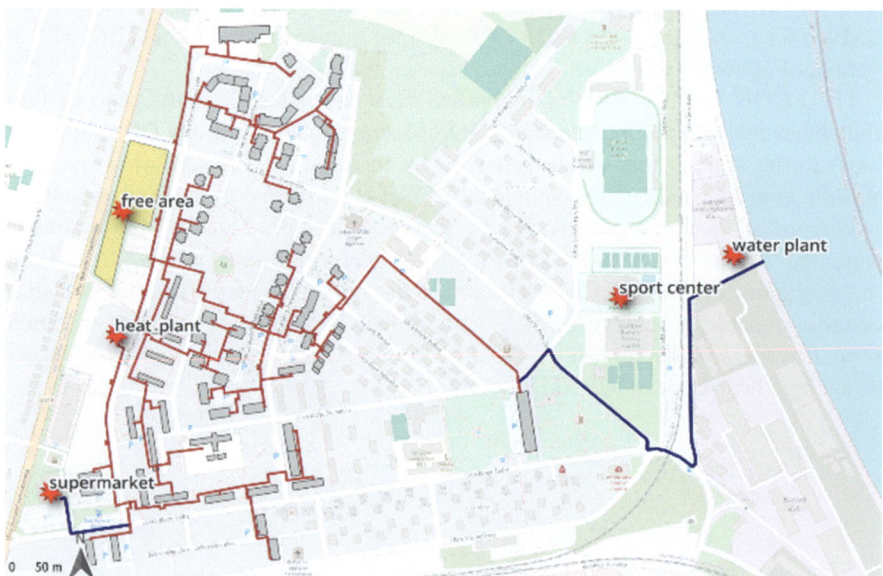

Fig. 1 Map of potential heat sources near the existing DH network and other points of interest. The red line indicates the current network path, while blue lines represent the possible network extensions required to connect new sources

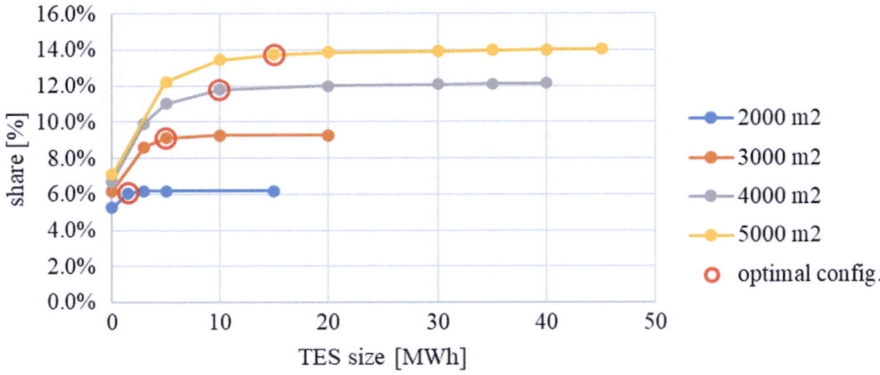

Fig. 2 Parametric analysis of solar share in the DH system for solar field expansion scenarios

an estimated annual availability of 1.2 GWh. An hourly typical availability profile was generated assuming continuous operation with diurnal and seasonal variations reflecting refrigeration load, resulting in a peak thermal power of 0.2 MW. The corresponding waste heat temperature, which is partially influenced by ambient air temperature, is assumed to remain above 25 °C based on other similar case studies. This results in an estimated COP of 3.65 and a corresponding HP capacity of about 0.3 MW.

As a first step of the pre-screening phase, an initial investigation focused on the potential expansion of the solar thermal field and the possible integration of a TES to maximize solar utilization. A parametric analysis was conducted to identify promising and cost-effective configurations, evaluating the overall area of the solar field (including the existing 850 m² installation). For the investigated total collector areas, Fig. 2 shows the evolution of the solar share in the DH system as a function of the TES capacity.

All curves exhibit an asymptotic trend, indicating that beyond a certain storage size, further increases in capacity result in negligible improvements in solar share, while investment costs continue to rise, making such configurations economically inefficient. The red-circled points highlight the optimal TES capacity (within the chosen parametric analysis) for each corresponding collector area, representing a trade-off between solar share maximization and investment cost. Table 2 presents the results for all the investigated total areas, for both cases without TES and those with the optimal TES capacity.

The integration of TES leads to a significant improvement in the actual solar utilization since it mitigates the mismatch between solar availability and heat demand. Despite the additional investment, storage enables more effective use of the solar resource by increasing the number of equivalent operating hours. While in some cases this results in a lower LCOH compared to the corresponding configurations without TES, in others the LCOH may slightly increase. For example, in the 2000 m² scenario, adding 1.5 MWh of TES raises the LCOH by around 3 €/MWh, but increases the solar share from 5.3% to 6.1%. Considering both the LCOH relative to

Table 2 Solar share and LCOH for different solar field configurations. TES volume can be derived assuming $\Delta T = 20$ K for Vukovar DH

	Without TES			With optimal TES			
Total area	solar share	eq. hours	LCOH	TES	Solar share	Eq. hours	LCOH
m²	–	h	€/MWh	MWh	–	H	€/MWh
2000	5.3%	518	62.0	1.5	6.1%	598	64.9
3000	6.2%	405	75.6	5	9.1%	599	71.2
4000	6.7%	330	90.1	10	11.8%	580	79.6
5000	7.1%	280	104.0	15	13.7%	541	88.5

the current DH tariff and the increase in renewable penetration, two configurations emerge as the most promising: 2000 m² of solar collectors with 1.5 MWh of TES, and 3000 m² with 5 MWh of TES.

Since solar thermal alone is not sufficient to meet the 50% RES + WH target, the pre-screening phase was completed with a parametric analysis on the addition of a HP to the two most promising solar + TES configurations. Different HP sizes were compared to identify scenarios that both achieve the decarbonization target and remain economically competitive with the DH tariff. In Fig. 3 the investigated configurations are plotted as a point in the space of HP equivalent hours and total renewable share (HP + solar). As expected, increasing HP size raises the RES share but reduces its equivalent hours.

Assuming an electricity price of 130 €/MWh—about 40% below the national average and achievable through a mix of grid supply, Power Purchase Agreements

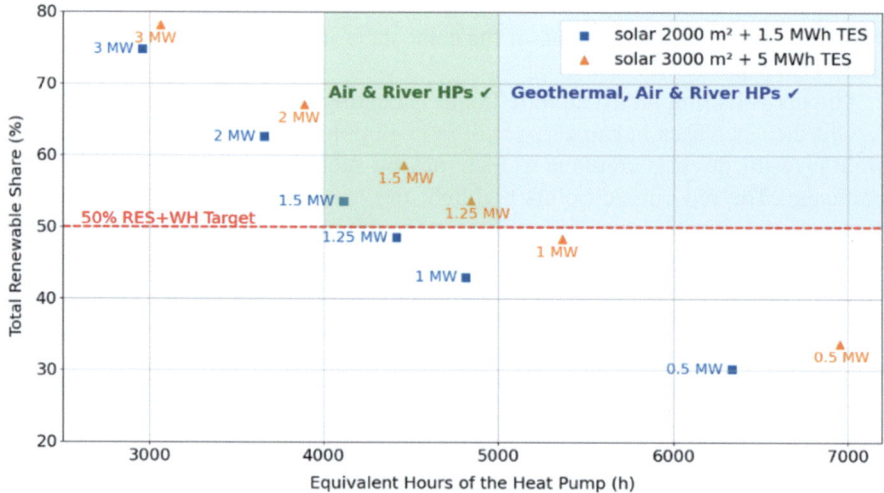

Fig. 3 Total renewable share and equivalent hours of the heat pump for different sizes of the heat pump

(PPAs), and on-site Photovoltaic (PV) self-consumption by the DH operator—the economic competitiveness of HPs was assessed. Considering estimated COP values and additional costs (connection costs for river HPs, boreholes installation costs for geothermal HPs), minimum equivalent operating hours of approx. 4000 are required for air and river HPs, and 5000 for geothermal HPs, to achieve a competitive LCOH with respect to the DH tariff. In Fig. 3, the shaded areas represent regions where configuration points must lie to meet both the 50% RES + WH target and the economic viability. Points outside these areas fail to meet one or both criteria and are thus excluded from further scenario simulations, effectively reducing the number of scenarios to be analyzed.

Due to the high cost of geothermal boreholes, no economically viable geothermal HP solution was identified. Instead, three cost-effective scenarios emerge for air and river HPs, with river-based options being particularly attractive, as the required new connection to the DH network could also be exploited to connect additional users. These three configurations represent the final output of the pre-screening phase and were selected for more detailed analysis:

- Scenario A: 2000 m^2 solar, 1.5 MWh TES, 1.5 MW river-HP
- Scenario B: 3000 m^2 solar, 5 MWh TES, 1.25 MW river-HP
- Scenario C: 3000 m^2 solar, 5 MWh TES, 1.5 MW river-HP.

The resulting equivalent LCOH of the new heat sources (i.e., accounting for the additional solar collectors, the storage, and the HP installation) is 81.51, 79.46, and 81.60 €/MWh for Scenarios A, B, and C, respectively. Scenario B emerged as the most promising, though with only a small cost advantage over the other two scenarios. Its daily heat share by sources is illustrated in Fig. 4. Note that HP is assumed to remain off throughout January for maintenance and to avoid operation in the coldest period, where the COP is lowest.

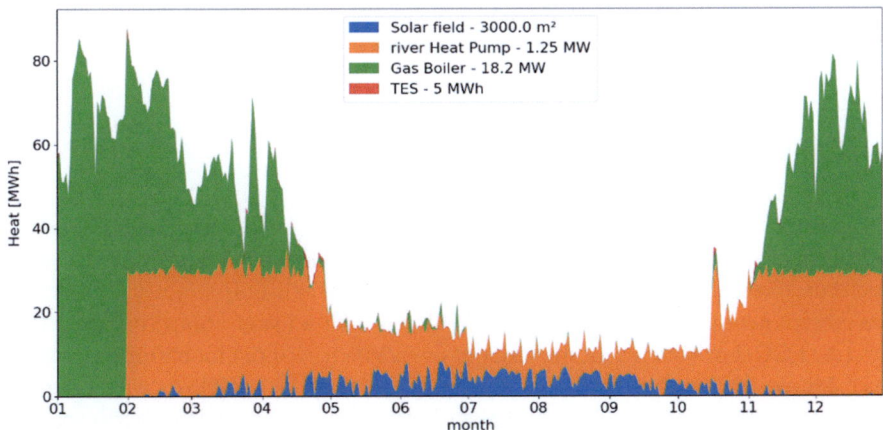

Fig. 4 Daily heat production by sources for scenario B

The heat production shares are 6.4% solar, 47.3% river heat pump, and 46.3% boilers. The heat production from RES accounts for 53.7% of the total, corresponding to about 6.8 GWh. Of this renewable production, 12.1% is generated by the 3000 m^2 solar field, while 87.9% comes from the river-HP.

The equivalent LCOH value of 79.46 €/MWh refers solely to the new installations (i.e., the additional solar collectors required to reach the total 3000 m^2, the storage, and the river heat pump). This equivalent LCOH can be interpreted as a weighted average between the LCOH of the new solar and TES system (101.57 €/MWh, with a 8.9% weight), and the LCOH of the river-HP (77.3 €/MWh, with a 91.1% weight). The increase in the solar and TES system LCOH compared to the previous analysis (71.2 €/MWh for 3000 m^2 solar and 5 MWh storage, see Table 2) results from the entire storage cost being allocated to the solar system, although the storage also serves the HP. Indeed, compared to the scenario without the heat pump, the overall solar share decreases from 9.1% to 6.4%. Regarding the heat pump, its LCOH of 77.3 €/MWh is composed of 25% investment costs (including equipment, electrical grid connection, and installation), 13% connection costs to the district heating network, 4% operation and maintenance costs, and 58% electricity purchase costs.

An alternative to Scenario B was also explored, involving the integration of a 0.3 MW HP recovering low-temperature WH from a supermarket, combined with a reduced 1 MW river HP to maintain roughly the same total HP capacity. In this Scenario D, despite the supermarket HP operating with a higher COP (3.65) compared to the river HP (2.92), the overall LCOH slightly increases compared to scenario B. This outcome arises because the LCOH of the river HP alone worsens as its capacity decreases from 1.25 MW to 1 MW. Indeed, while the HP installation costs scale down with capacity, the connection pipeline costs do not, since the pipeline must maintain the same diameter (DN 125, estimated to correspond to a total cost of about 900 €/m) for both sizes due to discrete diameter options. The next smaller diameter (DN 100, estimated to correspond to a total cost of about 780 €/m), with a transmission capacity limited to 0.8 MW, is insufficient for the 1 MW HP, preventing any cost savings on the connection.

4 Conclusions

The study presented in the paper for the DH of Vukovar allowed identifying a configuration able to reach the 2035 efficient DHC targets according to the EU EED, though just at the threshold of feasibility (LCOH about 79 €/MWh, only slightly smaller than the current heat tariff of about 82 €/MWh) and with a lower competitiveness than gas (the LCOH of which is of the order of 66 €/MWh, without including any carbon tax). Moreover, a rather favorable electricity price of 130 €/MWh was assumed, which requires a careful mix of grid electricity, PPAs, and/or PV. Scenario analysis showed that the identified configuration, featuring a 3000 m^2 solar field, a 1.25 MW river heat pump, and a 5 MWh thermal storage unit, is the most cost-effective among the options investigated (excluding deep geothermal or biomass, not considered in the study). With the inclusion of incentives expected to become available soon in Croatia, the proposed scenario would become more economically attractive. Finally, more advanced control solutions than the fixed priority rules used here could reduce operating costs and improve viability, for example through an optimizer dynamically determining dispatch order and source modulation.

From a methodological standpoint, the approach combines a source pre-screening phase with a scenario evaluation based on hourly energy balances. In the case of small-scale applications like Vukovar, this allows for an almost exhaustive evaluation of the limited number of possible combinations. However, when dealing with larger heat source portfolios or more complex systems, a full scenario-by-scenario simulation becomes impractical even after pre-screening. In such cases, integrating an optimization procedure after the pre-screening phase becomes necessary to efficiently explore the reduced solution space. Nevertheless, pre-screening remains valuable even in the presence of an optimizer, as it significantly reduces computational efforts by discarding clearly unfeasible options.

The resulting scenarios of course rely on aggregated calculations. The hydraulic feasibility of integrating new heat sources cannot be assessed in this way and would require more detailed thermo-hydraulic analysis and local studies to confirm assumptions. This will be part of the continuation of the HeatMineDH project, once the identified scenarios will be confirmed by a dedicated market investigation.

Despite the mentioned limitations, the described methodology offers a structured approach linking spatial mapping, individual technology evaluation, and scenario analysis. It enables the identification of the most competitive scenarios in the early planning phase of DH decarbonization and showed promising options for the considered case of Vukovar.

References

1. Lund R, Persson U (2016) Mapping of potential heat sources for heat pumps for district heating in Denmark. Energy 110:129–138. https://doi.org/10.1016/j.energy.2015.12.127
2. Nielsen S, Hansen K, Lund R, Moreno D (2020) Unconventional excess heat sources for district heating in a national energy system context. Energies 13:5068. https://doi.org/10.3390/en13195068
3. Dénarié A, Fattori F, Spirito G, Macchi S, Cirillo VF, Motta M, Persson U (2021) Assessment of waste and renewable heat recovery in DH through GIS mapping: The national potential in Italy. Smart Energy 1:100008. https://doi.org/10.1016/j.segy.2021.100008
4. Su C, Dalgren J, Palm B (2021) High-resolution mapping of the clean heat sources for district heating in Stockholm City. Energy Convers Manage 235:113983. https://doi.org/10.1016/j.enconman.2021.113983
5. Manz P, Billerbeck A, Kök A, Fallahnejad M, Fleiter T, Kranzl L, Braungardt S, Eichhammer W (2024) Spatial analysis of renewable and excess heat potentials for climate-neutral district heating in Europe. Renewable Energy 224:120111. https://doi.org/10.1016/j.renene.2024.120111
6. Pakere I, Feofilovs M, Lepiksaar K, Vītoliņš V, Blumberga D (2023) Multi-source district heating system full decarbonization strategies: Technical, economic, and environmental assessment. Energy 285:129296. https://doi.org/10.1016/j.energy.2023.129296

7. Daugavietis JE, Ziemele J (2024) District heating system's development decarbonization strategy assessment by system dynamics modeling and multi-criteria analysis. J Environ Manage 356:120683. https://doi.org/10.1016/j.jenvman.2024.120683
8. Dalle Nogare G, Cozzini M (n.d.) Report on the GIS tool for waste heat recovery opportunities individuation—Action C.6.4—LIFE4HeatRecovery
9. Macenić M, Kurevija T, Strpić K (n.d.) Systematic review of research and utilization of shallow geothermal energy in Croatia. https://doi.org/10.17794/rgn.2018.5.4
10. Danube HIS—Station HR5070_HYDRO VUKOVAR (n.d.). https://www.danubehis.org/results/HR5070_HYDRO.
11. Banks D (2012) An introduction to thermogeology: ground source heating and cooling, 2nd ed. Wiley-Blackwell, Chichester, West Sussex Somerset, N.J. https://doi.org/10.1002/9781118447512
12. Gueymard CA (2008) From global horizontal to global tilted irradiance: how accurate are solar energy engineering predictions in practice? San Diego
13. NASA POWER|DAV (n.d.). https://power.larc.nasa.gov/data-access-viewer/ (accessed 3 June 2005)
14. Calixto S, Köseoğlu C, Cozzini M, Manzolini G (2021) Monitoring and aggregate modelling of an existing neutral temperature district heating network. Energy Rep 7:140–149. https://doi.org/10.1016/j.egyr.2021.08.162
15. Agora Energiewende (n.d.) The roll-out of large-scale heat pumps in Germany. https://www.agora-energiewende.org/publications/the-roll-out-of-large-scale-heat-pumps-in-germany (accessed 3 June 2005)
16. Eurostat—Comparative price levels for investment, (n.d.). https://ec.europa.eu/eurostat/statistics-explained/index.php?title=Comparative_price_levels_for_investment
17. Lu Q, Narsilio GA, Aditya GR, Johnston IW (2017) Cost and performance data for residential buildings fitted with GSHP systems in Melbourne Australia. Data Brief 12:9–12. https://doi.org/10.1016/j.dib.2017.03.028
18. EUROSTAT—Database (n.d.)
19. Tehnostan—Vukovar District Heating tariffs (n.d.). https://tehnostan-vukovar.hr/toplinarstvo/cjenik/.

Open Access This chapter is licensed under the terms of the Creative Commons Attribution 4.0 International License (http://creativecommons.org/licenses/by/4.0/), which permits use, sharing, adaptation, distribution and reproduction in any medium or format, as long as you give appropriate credit to the original author(s) and the source, provide a link to the Creative Commons license and indicate if changes were made.

The images or other third party material in this chapter are included in the chapter's Creative Commons license, unless indicated otherwise in a credit line to the material. If material is not included in the chapter's Creative Commons license and your intended use is not permitted by statutory regulation or exceeds the permitted use, you will need to obtain permission directly from the copyright holder.

Exploiting Synergies of Data-Driven and Model-Based Approaches for Leakage Localization in District Heating Networks: Application of Improved Approaches

Dennis Pierl, Julia Koltermann, Kai Vahldiek, Bernd Rüger, Kai Michels, Andreas Nürnberger, and Frank Klawonn

Abstract Leakage detection and localization in District Heating Networks (DHNs) remains critical to maintain operational reliability and minimize economic and energy losses. Three different data-driven and model-based approaches have been proposed to solve this problem and delivered promising results: an approach for detecting and evaluating leakage-induced pressure waves (PWD), a numerical-analytical approach based on a district heating network model (MBSE) and a purely data-driven approach (ML). All these approaches rely on current measurement data from the network, i.e. pressure, flow rate and temperature. These approaches have been continuously improved ever since. The MBSE approach, which was previously based on a purely hydraulic DHN model, has been extended to include the thermal model equations, which allows better consideration of the available temperature measurement data. This temperature measurement data is also used by the ML approach in order to estimate the resulting potential for improvement with better preselections. These approaches are applied to the same historical measurement data of real DHN leakage events used in a previous study to evaluate the performance enhancements. First, the approaches are evaluated independently to quantify their individual improvements. Subsequently, as previously demonstrated, their interoperability is examined to exploit potential synergies to narrow down their search space and effectively locate

---

B. Rüger
SWM Services GmbH, München, Germany

D. Pierl (✉) · K. Michels
Institute of Automation Technology (University of Bremen), Bremen, Germany
e-mail: pierl@iat.uni-bremen.de

J. Koltermann · A. Nürnberger
Institute for Technical and Business Information Systems (Otto Von Guericke University), Magdeburg, Germany

K. Vahldiek · F. Klawonn
Institute for Information Engineering (Ostfalia University of Applied Sciences), Wolfenbüttel, Germany

© The Author(s) 2026
D. Vanhoudt (ed.), *Proceedings of the 19th International Symposium on District Heating and Cooling*, Lecture Notes in Networks and Systems 1700,
https://doi.org/10.1007/978-3-032-09844-3_13

leakages. The results are compared to the baseline established in the aforementioned study, highlighting the impact of the methodological extensions on the overall leakage localization performance. By combining the refined individual results of the three approaches, this study not only emphasizes the respective strengths of each method, but also underlines the importance of combining their capabilities to achieve a robust and efficient leakage localization framework for DHNs.

Keywords Leakage localization · Pressure wave evaluation · Machine learning · Numerical-analytical network model

1 Introduction

Leakages pose serious technical and economic risks for the operation of District Heating Networks (DHNs). In this context, there is an urgent need for procedures that immediately and centrally provide the operator with information about location and extent of a leakage—ideally in real-time and integrated into the existing process management system. Such procedures could provide systematic countermeasures as soon as possible, e.g. by automatically detaching affected network areas in order to minimize consequential damage and maintain network operation. In this regard, research focuses on approaches for leakage detection and localization based on current measurement data from the underlying DHN typically available—in particular pressure, temperature and flow measurement data. Considering these demands, a large variety of different model-based and purely data-based methods already exists, which are briefly described below.

In recent publications, the potential of hybrid methods combining physical models (Digital Twin (DT)) and data-based machine learning approaches has been increasingly emphasized. An approach of this kind was proposed by Yang et al. [5]. Here, a Neural Network (NN) was trained based on synthetic leakage data sets generated by a calibrated network model of the underlying DHN. The results achieved illustrate the high potential of such hybrid approaches. Zheng et al. [6] propose a model-based approach for long-distance district heating pipelines relying on a transient hydraulic model to simulate pressure wave propagation. The leakage location is determined applying Particle Swarm Optimization (PSO) with a final accuracy of < 480 m. A DT of a DHN was as well used by Zheng et al. [7]. PSO was applied for estimation and prediction of the heat demand within the DHN, resulting in a drastically improved modelling accuracy regarding pressures and return temperatures by up to 80%. Traditional reliability and asset management methods are increasingly considered inadequate for modern DHNs with highly available measurement data. As a result, recent research focuses on advanced, data-driven approaches and machine learning techniques to detect, assess, and even predict leakages and other faults in real time [3]. Approaches for water distribution networks (WDNs) can also be adapted, e.g. Basnet et al. [2] use machine learning models with simulated data for WDNs and highlight the challenges that arise from complex network structures like operational changes and uncertainties. In [1], features for the machine learning algorithm are extracted using wavelet decomposition, making the approach more robust to changing conditions, which also poses challenges in our work.

It is evident that the localization of leakages in DHNs remains a highly relevant field of research. Its continued relevance stems from the growing importance of a safe and sustainable heat supply in urban infrastructures. The approach pursued in this work combines both model- and data-based techniques and addresses different phases of leakages: transient effects, resulting from occurring pressure waves, as well as stationary effects, manifesting in altered pressure, flow and temperature measurement values. This combination provides a robust approach to detection and localization of leakages under realistic conditions as already shown in a case study based on measurement data from a real DHN [4]. From this work, the following research question arises: To what extent do refinements to the existing individual approaches improve accuracy and robustness of the combined (synergistic) approach to leakage localization in DHNs?

2 Methodology

This work focuses on three different approaches to leakage localization—Pressure Wave Detection (PWD), Model-based State Estimation (MBSE) and Machine Learning (ML)—presented in a previous contribution and examined and evaluated based on a case study [4]. These approaches have been continuously developed individually. The objective of this work is to discuss the impact of these improvements on the performance compared to the original state of development. The utilization of synergy effects has not changed and is adopted from the previous study: PWD essentially provides a preselection of five potential exclusion areas (EAs), subsequently used by MBSE and ML to narrow down their search space. Additionally, MBSE simulation results are used to substitute missing measurement data, providing synthetic data for ML.

2.1 Individual Approach Improvements

Only minor modifications have been applied to the PWD approach since the last study, which have no or only a marginal effect on its performance in terms of leak localization. Therefore, the results achieved by PWD are not being discussed here.

The MBSE approach, originally based on a purely hydraulic network model, is now extended to include stationary thermal model equations. In the original variant, the network state estimation only included the determination of the pressure and flow distribution in the DHN. The existing temperature measurement values were used solely to determine parameters of the transport medium such as density or viscosity. Now, the extended MBSE approach is based on a thermo-hydraulic network model which is capable of determining the temperature distribution throughout the network in addition to pressure and flow as part of the network state estimation. Although this significantly increases the number of state variables to be computed, the additional consideration of thermal effects provides a more precise mapping of real energy flows and an improved estimation of unknown consumption mass flows. In addition, the temperature measurement values available in the network are now systematically included in the MBSE approach, which further increases the consistency and validity of the estimation.

The ML approach localizes leakages solely based on stationary measurement data using a pre-trained NN for predicting the affected EA. Training of the underlying NN relies on synthetic data generated by the network model part of MBSE. Therefore, ML benefits from more accurate training data resulting from a more realistic thermo-hydraulic state estimation by MBSE and better preselections when using synergies.

3 Results

To evaluate the impact of the modifications to the individual approaches resulting in altered performance of leakage localization, data sets for 18 events within the real DHN (accumulated pipe length of approx. 190 km, divided into 28 EAs) are assessed. The events' locations within the DHN topology are indicated in Fig. 1 and their corresponding total leakage amount $\dot{m}_L$ the rate of increase (gradient) of the leakage amount $\Delta \dot{m}_L / \Delta t$ are mentioned in Table 1. The performance of each approach is determined by the performance criterion (PC, ranging from 0 to 100%)

$$PC = \frac{N - r}{N - 1} \cdot 100\,\% \tag{1}$$

where r is the rank of the actually affected EA in an EA ranking list generated by the approach for an event and N is the total number of EAs within the DHN.

3.1 Individual Results

First, the performance of the individual modified approaches for leakage localization is outlined on the basis of the 18 events. For the PWD approach, there are no significant changes in the results compared to the previous analyses as shown in [4].

However, considerable improvements can be seen in the performance of the modified MBSE approach. The extension of the approach and the additional use of the temperature measurement values led to a significant increase in the overall PC from formerly 73.3–84.6%. In particular, the actually affected EA is now found more frequently at the very top of the EA ranking list for leakage localization: In the

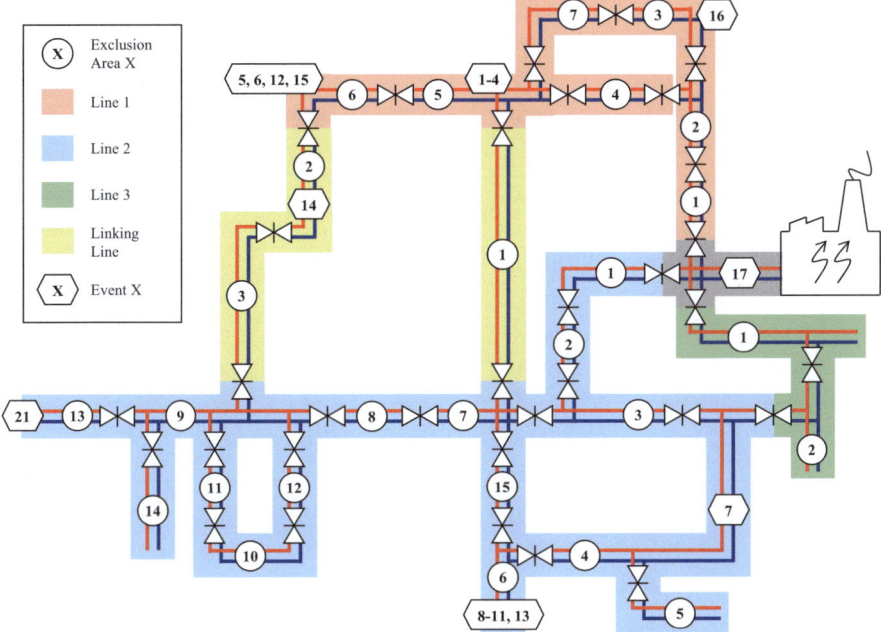

Fig. 1 Simplified overview of the topology of the DHN under investigation including indications for the locations of all leakage events according to Table 1

present case study, it is now among the top three ranks in 12 out of the total of 18 events, whereas this was only the case for 7 of the events in the previous study. The worst ranking of the correct EA has also improved noticeably—while it was previously in 11th place, it has now been reduced to the 6th place in the EA ranking list. As part of the development of MBSE, though, its high sensitivity with respect to its thermal model parameters emerged. In particular, the heat transfer coefficient h of the pipes turned out to be a critical impact factor for the approach's reliability. A supplementary parameter study was therefore conducted to investigate the influence of deviations in the heat transfer coefficient h on the quality of the results. It was found that even small deviations of $\pm 2\%$ around its nominal value led to considerable deterioration in the correct allocation rate. As a result, incorrect localizations occur more frequently, as the leakage probabilities determined by the approach for each EA, on which the EA rankings are based, are shifted. This effect is shown in Fig. 2 exemplarily for event #4 (EA designation: Line X, Exclusion Area YY—*LX-EAYY*).

Compared to the previous results, the ML approach shows a minor improvement in its individual leakage localization results when using training data generated on the basis of the revised thermos-hydraulic network model of MBSE. The mean PC over all events increases from originally 55.3% to now 56.8%. In addition, the standard deviation in the results has reduced from previously 17.4% to just 15.0%, which

Table 1 Comparison of the localization results for 18 events in a real DHN applying the improved approaches both individually and in combination to exploit synergies

#	$\dot{m}_L$ [m³/h]	$\Delta\dot{m}_L/\Delta t$ [m³/h/min]	MBSE (PC) indiv.	comb.	ML (Ø PC ± STD) indiv.	comb.
1	78.0	36.3	92.6%	100.0%	37.4 ± 21.9%	91.5 ± 4.6%
2	146.1	39.0	100.0%	100.0%	79.6 ± 13.3%	90.4 ± 5.0%
3	221.4	100.3	100.0%	100.0%	63.7 ± 14.7%	91.1 ± 6.6%
4	251.9	40.8	100.0%	100.0%	75.8 ± 12.8%	94.4 ± 5.8%
5	94.0	16.7	100.0%	0.0%	74.1 ± 12.8%	95.6 ± 5.2%
6	113.5	9.0	96.3%	100.0%	65.3 ± 18.7%	94.0 ± 4.6%
7	100.0	35.2	0.0%	0.0%	78.8 ± 14.4%	–
8	50.0	25.7	92.6%	92.6%	48.1 ± 25.6%	93.3 ± 3.8%
9	100.0	27.0	88.9%	0.0%	63.0 ± 15.4%	–
10	124.0	68.6	88.9%	100.0%	47.0 ± 17.5%	–
11	158.0	56.7	100.0%	100.0%	74.6 ± 13.1%	93.2 ± 4.1%
12	140.0	32.9	100.0%	100.0%	55.9 ± 18.0%	94.2 ± 4.7%
13	231.7	155.6	96.3%	96.3%	41.1 ± 15.9%	94.8 ± 3.0%
14	76.9	10.6	88.9%	–	59.8 ± 12.2%	-
15	195.6	65.7	96.3%	96.3%	37.7 ± 13.3%	90.5 ± 5.6%
16	158.4	122.6	100.0%	100.0%	59.8 ± 11.3%	94.3 ± 3.7%
17	421.3	3530.0	81.5%	88.9%	–	–
21	133.0	234.5	0.0%	–	4.0 ± 0.0%	–
Ø			84.6%	79,6%	56.8 ± 15.0%	± 4.7%

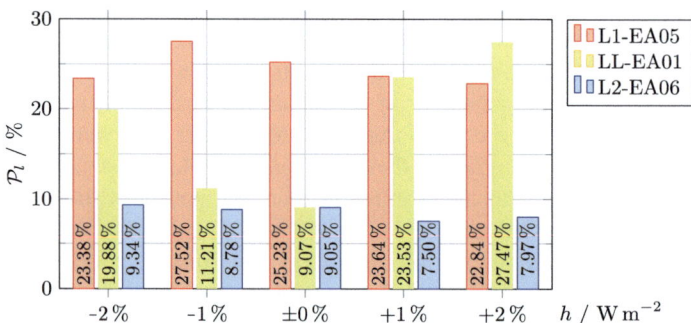

Fig. 2 Dependency of the leakage probabilities P_l in the exclusion area ranking (top3) for varying heat transfer coefficients h of event #4 (actually affected exclusion area: L1-EA05)

suggest that the localization is now more robust against effects such as measurement noise.

3.2 Combined Approach Results

Below, the results achieved by the MBSE and ML approaches are outlined, exploiting synergy effects such as prior knowledge of the damage location provided by PWD. The associated PC values achieved in this context are as well listed in Table 1.

When using synergy effects, the MBSE approach shows only a slight increase in the PC from originally 78.9% to now 79.6% compared to the solely hydraulic model [4]. There even occurs a deterioration in the PC with synergies compared to its individual application. This is mainly caused by the results for events #7 and #9, where the leakage localization fails entirely in the combined case. In case of event #5, an EA neighboring the EA actually affected (*L1-EA06*), namely *L1-EA05*, is classified with a leakage probability of 100.0% and therefore ranked 1st on the ranking list. All other EAs therefore exhibit a leakage probability of 0.0% and are thus assigned a rank of $r = 28$, resulting in a PC of 0.0%. For the events #14 and #17, the solver does not converge, similar to the previous study, so that no statement can be made here. For all other events, the combined approach shows equivalent or better values of the PC compared to the individual application of MBSE.

Using a preselection of the most possibly affected EAs is also beneficial for the ML approach. Here, a significant increase in the mean PC from 56.8% to 93.1% and an even further reduction in the standard deviation of the PC from 15.0% to 4.7% can be observed. Compared to the original results, however, there is no noticeable change when using synergy effects. Instead, there is even a slight deterioration in the results, both in terms of mean PC as well as its standard deviation (originally 93.9 ± 3.6%).

4 Discussion

Overall, the analyses carried out demonstrate that extending the underlying network model of MBSE to include thermal components can significantly improve both the accuracy and reliability with regard to leakage localization. However, the high sensitivity of MBSE with respect to the thermal model parameters indicates that highly accurate model calibration is necessary to achieve high reliability. This involves the precise modelling of the topology of the pipe network (as in case of the solely hydraulic model), as well as a comprehensive measurement database for model calibration.

Concerning the exploitation of synergy effects among the approaches, it can be said that although they provide several advantages, this must also be reviewed critically depending on the situation. In most of the cases analyzed (13 out of 18 for

MBSE and 12 out of 18 for ML), it was found that if the approaches already achieved good results individually, their combination then confirmed or even improved these results. However, if one approach yields very poor results (e.g. events #9 and #14), this may lead to the other methods being negatively affected. If the actually affected EA was included in the PWD's preselection for the other approaches for events #9 and #14, this would lead to a significant improvement in the overall result. In this case, MBSE would hypothetically achieve a PC of 100.0% for event #9 and a PC of 96.3% for event #14, resulting in a mean PC across all events of approx. 86.5%. This would subsequently lead to a further improvement in the results compared to its individual application. For ML, the similar applies. If the actually affected EA hypothetically was included in the preselection, the approach would not fail and results with a high mean PC of approx. 90% on average and low standard deviations of approx. 5% would be obtained for the events #7 #9, #10, #14 and #21.

Overall, the results achieved in the present case study show that individual enhancements carried out to the approaches still have potential for optimization, depending on the approach. By exploiting synergy effects between the individual approaches, the performance with regard to leakage localization could be increased even more. However, it could also be observed that the currently applied (binary) information exchange between the approaches is most likely not the best solution. Here, further optimization potential remains for the combined approach.

5 Conclusion

In the present work, three existing approaches for leakage localization in District Heating Networks—Pressure Wave Detection, Model-based State Estimation, Machine Learning—are further investigated. The approaches have been continuously improved and methodically enhanced. The aim was to analyze the impact of these enhancements on the approaches individually, but also with regard to the synergy exploitation effects in terms of leakage localization. The results achieved can be summarized as follows:

- The MBSE approach was extended by a stationary thermal model to consider available measured temperatures across the network. This extension led to a significantly improved localization performance. Nevertheless, the method showed high sensitivity to the thermal model parameters, requiring a carefully calibrated network model.
- The ML approach individually exhibits minor improvements due to more accurate training data, again emphasizing the impact of the quality of the training data on the robustness of the prediction. However, it turns out that ML benefits much less from the thermal extension of the underlying network model compared to MBSE.
- Exploiting synergy effects leads to a significant increase in the correct allocation rate of leakages for both MBSE and ML. The use of the thermo-hydraulic model to generate more accurate synthetic training data leads to moderate improvements

in the ML approach. However, it is evident that the current approach to exploit synergies using a binary preselection of possibly affected EA not yet unfolds its full potential. The challenges already addressed in the preceding study still remain. Consequently, further optimization to the individual approaches hardly pose any potential for improvement. Instead, the way synergy effects are used needs to be reflected.
- Further steps for future research on this topic could include: Implementation and extensive testing of ML and MBSE in online operation (PWD already implemented) using real-time measurement data; Refinement of the synergy exploitation strategies, particularly by replacing the current binary filtering of potential EAs with a weighted consideration within the objective functions of the individual approaches; Exploring self-learning/continuous learning approaches for ML, allowing to improve over time by incorporating newly acquired measurement data; Narrowing down the leakage to supply/return network by PWD and providing this information to ML and MBSE.

Overall, the study carried out shows that the modifications made to the approaches have significantly improved the performance in terms of leakage localization.

Acknowledgements We'd like to thank the Federal Ministry of Economic Affairs and Climate Actions (BMWK) and the Project Management Jülich (PtJ) for funding the project as part of the Federal Government's 7th Energy Research Program (funding code: 03ET1624 A–D). Furthermore, we'd like to thank Mr. Christian Lukas (Development Office for Physical Technology) for his dedicated and competent support in network modelling in STANET as well as Mr. Friedrich Fischer-Uhrig (Fischer-Uhrig Engineering GmbH) for providing STANET for academic purpose.

Supported by:

on the basis of a decision
by the German Bundestag

References

1. Adraoui M, Azmi R, Chenal J, Diop EB, Abdem SAE, Serbouti I, Hlal M, Bounabi M (2024) A two-phase approach for leak detection and localization in water distribution systems using wavelet decomposition and machine learning. In: Computers & Industrial Engineering, vol 197, p 110534
2. Basnet L, Ranjithan R, Mahinthakumar K (2023) Supervised learning approaches for leak localization in water distribution systems: impact of complexities of leak characteristics. J Water Resour Plan Manag 149:8

3. Losi E, Manservigi L, Spina PR, Venturini M (2024) Data-driven approach for the detection of faults in district heating networks. In: Sustainable energy, grids and networks, vol 38, p 101355
4. Pierl D, Vahldiek K, Koltermann J, Rüger B, Michels K, Klawonn F, Nürnberger A (2023) Exploiting synergies of data-driven and model-based approaches for leakage localization in district heating networks. In: 18th international symposium on district heating and cooling, Beijing (China)
5. Yang G, Xing D, Wang H (2024) Leak localization in District Heating Networks integrating physical model-based and data driven-based methods: Impact of dataset construction on model performance. Energy 308:132839
6. Zheng X, Hu F, Wang Y, Zheng L, Gao X, Zhang H, You S, Xu B (2021) Leak detection of long-distance district heating pipeline: A hydraulic transient model-based approach. Energy 237:121604
7. Zheng X, Shi Z, Wang Y, Zhan H, Tang Z (2024) Digital twin modeling for district heating network based on hydraulic resistance identification and heat load prediction. Energy 228:129726

Open Access This chapter is licensed under the terms of the Creative Commons Attribution 4.0 International License (http://creativecommons.org/licenses/by/4.0/), which permits use, sharing, adaptation, distribution and reproduction in any medium or format, as long as you give appropriate credit to the original author(s) and the source, provide a link to the Creative Commons license and indicate if changes were made.

The images or other third party material in this chapter are included in the chapter's Creative Commons license, unless indicated otherwise in a credit line to the material. If material is not included in the chapter's Creative Commons license and your intended use is not permitted by statutory regulation or exceeds the permitted use, you will need to obtain permission directly from the copyright holder.

Data-Driven Fault Detection and Diagnosis in District Heating Substations and the Impact of Return Temperature Reduction

Vera Alieva, Tiedo Behrends, Vera Boß, Peter Stange, and Clemens Felsmann

Abstract Reducing return temperatures in district heating (DH) networks is crucial for improving overall efficiency, minimising thermal losses, and enabling the integration of renewable energy sources. Achieving this requires a detailed understanding of DH substations performance, as faults or inefficiencies in substations are a common cause of elevated return temperatures. This case study introduces a data-driven approach for fault detection and diagnosis (FDD) in DH substations, aiming to identify such inefficiencies and support targeted performance improvements. The methodology is implemented in a Python-based tool that analyses and visualises smart meter data collected from the primary side of substations. The tool offers geospatial mapping, performance indicators (e.g., return temperature, hydraulic capacity utilization), and metadata filtering (e.g., building type). Using this approach, four recurring fault patterns are identified. To support fault handling prioritisation, the tool segments inefficient substations through scatterplot visualisations. The proposed approach offers utility companies a means to enhance system reliability, reduce emissions, and improve operational performance by systematically identifying and addressing performance issues.

Keywords Fault detection and diagnosis · Return temperature optimization · Anomaly detection · Network performance

1 Introduction

DH networks play a key role in sustainable urban energy systems, but their efficiency heavily depends on the performance of individual customer substations. Previous studies indicate that a significant share of substations operate sub-optimally; for example, 50–60% of substations in some residential dwelling groups were found

---

V. Alieva (✉) · T. Behrends · V. Boß · P. Stange · C. Felsmann
Chair of Building Energy Systems and Heat Supply, Institute of Power Engineering, Technical University of Dresden, Dresden, Germany
e-mail: vera.alieva@tu-dresden.de
URL: https://tu-dresden.de/ing/maschinenwesen/iet/gewv

to have faults or errors that lead to inefficient operation (e.g., abnormally high return temperatures) [1]. Such inefficiencies at the substation level propagate through the network, causing higher return temperatures that degrade overall system performance. High return temperatures increase heat losses in distribution pipes and reduce generation efficiency, especially for modern low-temperature or renewable heat sources. They can also compel higher volume flow rates (to deliver the same heat with a smaller temperature difference), wasting pumping energy and potentially limiting network capacity [2, 3]. These factors underscore that poorly performing substations are a barrier to energy efficiency and decarbonization goals. Traditional methods for identifying underperforming substations have relied on manual inspections or reactive responses to customer complaints, which are labor-intensive and often overlook hidden faults. Many utility companies currently lack systematic methods to pinpoint poorly performing substations [4]. However, the increasing deployment of smart heat meters and digital monitoring offers an opportunity to change this. Smart meter data, typically collected for billing, can be used for continuous performance analytics and early fault detection. Simple key performance indicators like the average return temperature or the temperature difference have been found effective metrics to identify substations with negative effects on network efficiency. Using such data-driven insights, utility companies can transition from reactive maintenance to a condition-based maintenance strategy, in which faults are detected and addressed proactively.

The evaluated DH network is divided into several subnetworks and supplies heat to around 1900 customers. It is characterised predominantly by high supply and return temperatures classifying it primarily as a second-generation network according to Dalla Rosa et al. [5]. As a result of the continuous roll-out of smart heat meters around 75% of all DH substations are now equipped accordingly. Historically, the customer structure has evolved so that a large proportion consists of small single- and multi-family dwellings. The next customer group comprises large residential complexes, supplemented by a number of municipal buildings and commercial properties. There is no available documentation on the substations configurations, as the ownership boundary is usually located immediately after of the shut-off valves, placing the substations within the domain of the customers.

2 Data Acquisition and Preprocessing

The smart heat meter data from the primary-side of the substation is transmitted via LoRaWAN technology and stored in a time-series database. These heat meters are mandatorily installed for billing purposes according to heating cost regulations, but in this context, they also provide high-resolution data. Occasionally, outages occurred for various reasons, resulting in high data losses. The standard data resolution is one hour, although in selected cases the transmission interval was adjusted to 15 min. The basis for identifying anomalous or inefficient behaviour is both the heat meter measurement data (measured supply ϑ_{sup} and return temperatures ϑ_{ret}, volume flow

$\dot{V}$, calculated thermal power $\dot{Q}$, aggregated thermal energy Q and volume V) and the available metadata (geographical location within the network, contractual connected load, design volume flow, network supply and return temperature, and building type). Design volume flow $\dot{V}_{\text{design}}$ is calculated based on the DH network temperatures specified in the Technical Connection Conditions (TCC) and the contractually agreed connected load of the substation.

The incoming data is initially checked for availability and plausibility. Scarce data availability (e.g. <10%) makes future evaluation impossible.

The plausibility of the measured data is automatically verified based on the following criteria:

1. Supply and return temperature constraints:

$$\vartheta_{\min} \leq \vartheta_{\text{sup}} \leq \vartheta_{\text{sup, network}} \ \wedge \ \vartheta_{\min} \leq \vartheta_{\text{ret}} \leq \vartheta_{\text{sup}}$$

Hereby, $\vartheta_{\min}$ is defined as the minimal ambient temperature in the area surrounding the substation, and $\vartheta_{\text{sup, network}}$ refers to the DH network supply temperature. During active operation of the substation, the supply temperature must exceed the return temperature: $\vartheta_{\text{sup}} > \vartheta_{\text{ret}}$

2. Volume flow constraints:

$$0 \text{ m}^3/\text{h} \leq \dot{V} \leq \dot{V}_{\text{design}} \cdot 1,1$$

Since it is not always possible to precisely set the flow limiter a tolerance of $\pm 10\%$ is permitted. Negative volume flow values are excluded, as the evaluated DH network does not include bidirectional substations.

3. Temperature-flow correlation:

$$\dot{V} = 0 \ \Rightarrow \ \frac{d\vartheta_{\text{sup}}}{dt} \leq 0$$

An increase in supply temperature must not be observed when the associated volume flow is zero.

4. Flow-temperature consistency:

$$\frac{d\dot{V}}{dt} > 0 \ \Rightarrow \ \left(\left| \frac{d\vartheta_{\text{sup}}}{dt} \right| > 0 \ \vee \ \left| \frac{d\vartheta_{\text{ret}}}{dt} \right| > 0 \right)$$

The volume flow must not increase without a corresponding change in supply or return temperatures.

3 Detection of Abnormal Behaviour

For further evaluation, only operating states in which the substation is active are considered. Periods of inactivity as well as startup and shutdown phases are filtered out ($\dot{V} > 0 \wedge \vartheta_{\text{sup}} \geq 0{,}8 \cdot \vartheta_{\text{sup, network}}$).

Based on the following conditions, key performance indicators are defined to identify substations as "anomalous" or inefficient. A substation is considered efficient if it shows high temperature difference with low volume flow.

1. **Supply Temperature Drop**: The cooling down of the supply temperature along the flow path is evaluated using the difference $\vartheta_{\text{sup, network}} - \vartheta_{\text{sup}}$. The evaluation threshold is defined based on the DH network supply temperature level. Potential causes for high deviation include hydraulic issues in the network or excessive thermal losses. Thermal transport delay must be taken into account and corrected for.
2. **Return Temperature**: The return temperature is assessed against the threshold values specified in the TCC. Both the average value and its frequency of exceedance are evaluated. If the criterion is met, the relative volume flow $\left(\dot{V}_{\text{rel}} = \frac{\dot{\tilde{V}}}{\dot{V}_{\text{design}}}\right)$, also referred to as hydraulic capacity utilisation, is used to determine whether the substation is frequently or predominantly operated at full capacity. In such case, this may indicate the following issues: insufficient return temperature limitation or lack thereof; heating curve calibration is too high; causes may lie on the secondary side of the substation (e.g., circulation pump operates continuously without regulation).
3. **Temperature Difference**: If the difference $\Delta\vartheta = \vartheta_{\text{sup}} - \vartheta_{\text{ret}}$ is less than $\Delta\vartheta_{\min}$ while the substation shows frequent operation, it can be inferred that no significant heat extraction takes place, yet high return temperatures are fed back into the network. Possible causes include: uncontrolled operation of the secondary-side circulation pump (as mentioned above); fouled heat exchanger; incorrect setpoints in the DH controller; faulty control valve. If the temperature spread exceeds $\Delta\vartheta_{\max}$ while $\dot{V}_{\text{rel}}$ indicates operation at full capacity, this may also signal improper system behaviour, potentially due to: incorrectly positioned temperature sensors; poorly insulated pipes; undersupply to the substation. The $\Delta\vartheta_{\min}$ and $\Delta\vartheta_{\max}$ thresholds result from the temperature differences in the summer and winter DH network operating modes, as specified in TCC.
4. **Correlation between thermal power and outdoor temperature**: The extrapolated heat load is derived via linear regression up to design outdoor temperature. The ratio between the contractually defined connected load and the extrapolated load is referred to as an **under-** or **over**dimensioning factor. This allows for conclusions on whether the connected load should be adjusted. Additionally, a consistent base load (outdoor temperature-independent demand) may indicate domestic hot water (DHW) heating via DH. The slope of the regression line can also provide indications about the construction period and/or state of renovation of the building.

Figure 1 shows the geospatial mapping in DH network of performance indicators along with information on data availability and measurements errors. It also includes the frequency distributions of return temperatures for various substations within Subnetwork 1 and for the building type large residential dwellings. The background map has been removed for anonymisation purposes.

4 Fault Pattern Identification

The key performance indicators help to detect substations abnormal behaviour, however, they do not always allow for a clear identification of the underlying causes. In numerous substations, similar behaviour patterns have been observed. To systematically capture these recurring patterns on the primary side of the district heating substation, so-called fault patterns have been defined. Each fault pattern is associated with potential causes of the observed behaviour. A total of four fault patterns have been identified and consecutively numbered. This numbering also serves as a guideline for prioritizing fault correction.

Fault Pattern 1 describes a scenario in which the volume flow rate remains constant while a high return temperature is fed back into the DH network. Potential causes for this behaviour include a defective DH control valve, improperly configured control parameters, an excessively high heating curve setting, or an inactive return temperature limitation.

Fault Pattern 2 is fundamentally similar to Fault Pattern 1 but occurs at a consistently low volume flow rate. This may be caused by a clogged or malfunctioning DH control valve or an inactive return temperature limitation.

Fault Pattern 3 describes a condition in which the volume flow rate never falls to zero (meaning the substation is always active). Possible causes include a defective or fouled control valve, or incorrectly positioned or faulty sensors. This evaluation is only relevant during non-heating periods. For certain substation configurations, such as those with DHW heating via flow-through principle, this behaviour does not necessarily indicate a fault. Therefore, the analysis result should be interpreted as a potential anomaly, which must be reviewed by the user, taking into consideration any available metadata regarding the substation configuration.

Fault Pattern 4 describes a state of significant fluctuations in volume flow rate, indicating hydraulic instability. Possible causes include improper sizing of DH control valve (e.g., due to oversizing), oversizing of the heat exchanger in indirectly connected systems, or unsuitable controller settings (P, I, D parameters).

In addition, **hydraulically critical** substations are identified. These are characterised by high hydraulic capacity utilisation in combination with low or optimal return temperatures—corresponding to a large temperature difference. In such cases, a reduction in supply temperature may potentially result in undersupply or insufficient heat delivery to the connected building systems.

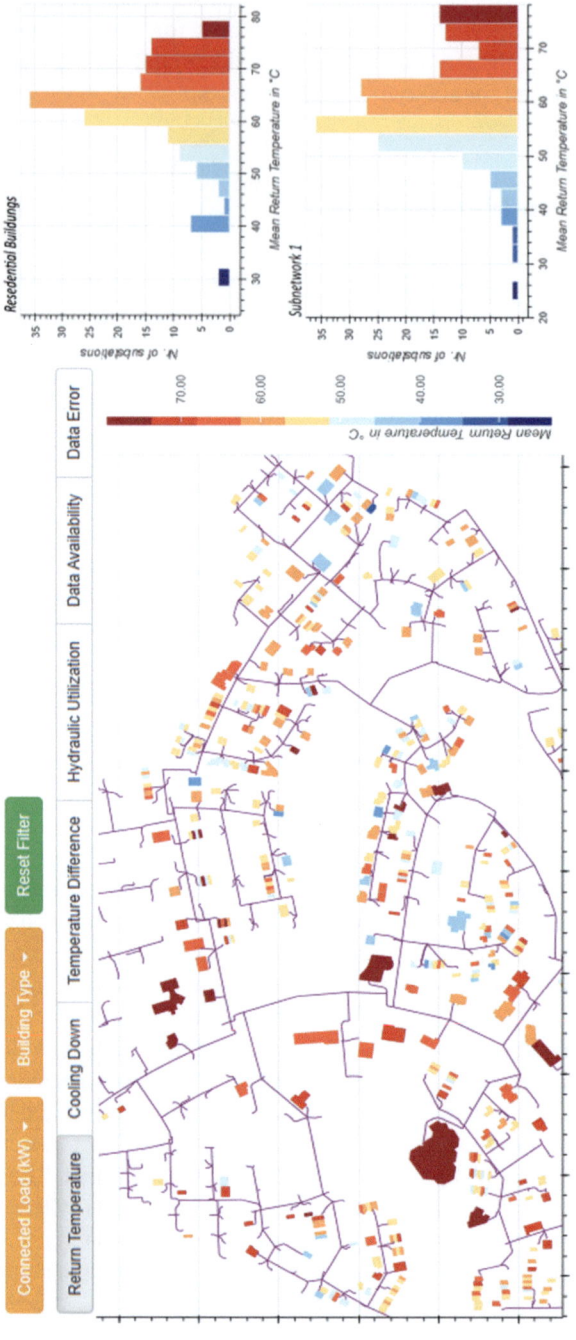

Fig. 1 Geospatial mapping of the substations KPIs

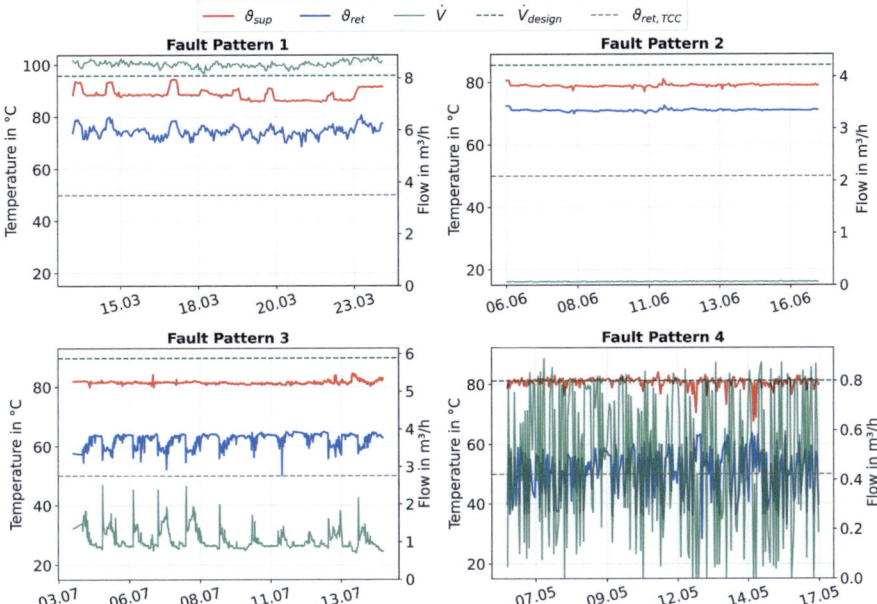

Fig. 2 Fault Patterns

Figure 2 illustrates the four fault patterns based on exemplary time series of supply and return temperatures as well as volume flow rate. Each plot represents a typical substation behaviour corresponding to one of the defined patterns.

Figure 3 illustrates the relationship between average return temperature and connected load for Fault Pattern 1. Each point represents a substation. For better interpretation, threshold values were defined, specifically, the TCC return temperature for a subnetwork and a connected load of 100 kW, dividing the plot into four mathematical quadrants. Substations located in the upper-right quadrant should be prioritized for further investigation and fault handling, as they have the greatest impact on the DH network's return temperature.

5 Return Temperature Reduction

Building upon the previously described methods for identifying inefficient DH substations behaviour, this evaluation quantifies the impact of fault mitigation by estimating the optimisation potential achievable through a reduction in return temperatures.

A reduction in return temperatures offers a range of easily accessible benefits for both the network operator and consumers. For the DH network operator, selected advantages include a decrease in volume flow at constant supply temperature, which lowers pumping power consumption, and the possibility of supply temperature reduc-

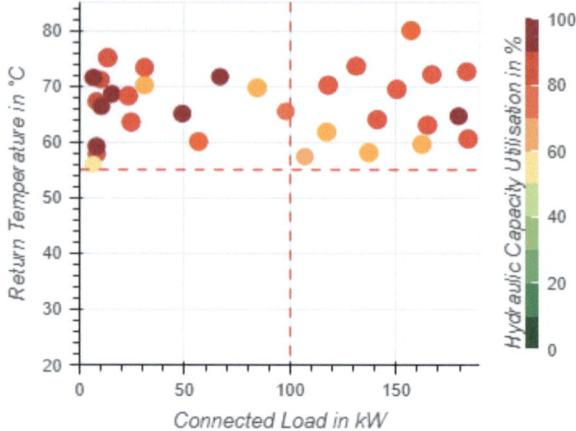

Fig. 3 Segmentation of substations with Fault Pattern 1

tion at constant volume flow, thereby improving the efficiency of renewable generators such as heat pumps. Furthermore, a higher temperature spread enables additional hydraulic capacity for connecting new consumers and contributes to lower thermal losses in the distribution network. On the consumer side, potential benefits include financial savings through bonus-malus incentives, reduced heat demand during summer (where no DHW is supplied via DH), enhanced operational reliability, and a decrease in CO_2 emissions.

The potential for reducing the DH networks return temperature is quantified by calculating the mixed return temperature based on the available metering data. Under the assumption that previously existing faults, which resulted in high return temperature breaching the TCC requirements, have been mitigated, and the return temperature now conforms to the TCC. Based on this assumption, the targeted mixed (i.e., optimal) return temperature is determined, along with the corresponding volume flow required to achieve this temperature, under the condition that the supply temperature remains unchanged. This is illustrated exemplarily in Fig. 4 for one of the subnetworks. Addressing faults in 55 of the 129 inefficient substations reduces the return temperature by around 12 K and lowers the circulated volume flow within the subnetwork by an average of 60%. With knowledge of the specific pump configuration, the potential energy and costs savings can be further quantified. This quantification approach proved effective for stakeholder communication.

It should be noted that the quality and reliability of this evaluation strongly depend on the available data—specifically, the data coverage and the number of available meters.

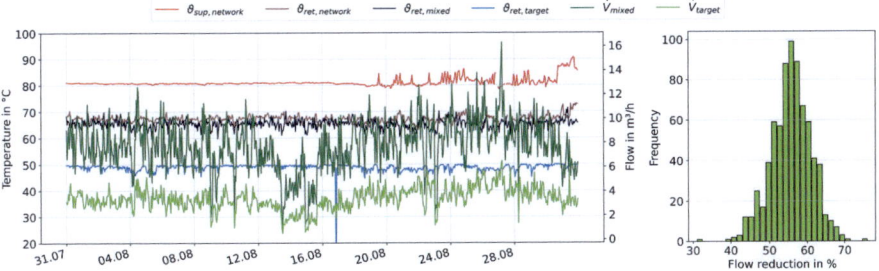

Fig. 4 Volume flow reduction due to return temperature reduction

6 Conclusion

This study demonstrates that automated, data-driven fault detection in DH substations is highly beneficial with simple metrics. By combining statistical diagnostics, geospatial analysis, and standardized fault pattern recognition, utility companies can substantially improve their network performance. Precise monitoring of substations enables the utility companies to identify substation operating inefficiently at an early stage and to proactively address the abnormal behaviour. The presented approach draws upon information made available through digitalisation and contributes directly to the decarbonisation of urban heating systems.

The actual process for mitigating substation faults is currently under development. A major challenge lies in the fact that substations are typically owned by the customers, which limits the utility company's direct intervention capacity. Consequently, effective fault correction will require suitable incentives and innovative business models to encourage customer cooperation. Nevertheless, leveraging the diagnostic results to inform and engage customers represents an essential first step towards operational improvement.

Future work will focus on extending the framework to include more fault patterns, characterisation of substations secondary-side based entirely only on the available primary-side data. At the same time a process-oriented evaluation that monitors the real-time effects of fault mitigation will occur. In addition, the impact of inefficient substations and specific fault types on neighbouring substations and the overall network will be investigated.

References

1. Gadd H, Werner S (2015) Fault detection in district heating substations. Appl Energy 157:51–59. https://doi.org/10.1016/j.apenergy.2015.07.061
2. Frederiksen S, Werner S (2013) District heating and cooling. Studentlitteratur AB, Lund
3. Paulick S, Boß V (2023) Schlussbericht zum Projekt EnEff:Wärme-ZellFlex: Identifikation urbaner Zellstrukturen für flexible Wärme- und Temperaturverteilung in Wärmenetzen

4. Månsson S, Thern M, Johansson Kallioniemi P-O, Sernhed K (2021) A fault handling process for faults in district heating customer installations. Energies. https://doi.org/10.3390/en14113169
5. Dalla Rosa A, Li H, Svendsen S, Werner S, Persson U, Rühling K, Felsmann C, Crane M, Burzynski R, Bevilacqua C (2014) IEA DHC Annex X report: toward 4th generation district heating: experience and potential of low-temperature district heating, IEA DHC|CHP

Open Access This chapter is licensed under the terms of the Creative Commons Attribution 4.0 International License (http://creativecommons.org/licenses/by/4.0/), which permits use, sharing, adaptation, distribution and reproduction in any medium or format, as long as you give appropriate credit to the original author(s) and the source, provide a link to the Creative Commons license and indicate if changes were made.

The images or other third party material in this chapter are included in the chapter's Creative Commons license, unless indicated otherwise in a credit line to the material. If material is not included in the chapter's Creative Commons license and your intended use is not permitted by statutory regulation or exceeds the permitted use, you will need to obtain permission directly from the copyright holder.

Data Pre-processing Methods Enhancing Heat Cost Allocator Measurement Usability

Qinjiang Yang, Fabio Saba, Marina Orio, Marco Santiano, Emanuele Audrito, Robbe Salenbien, and Michele Tunzi

Abstract Heat cost allocators (HCAs) are devices mounted on radiators to fairly allocate heating consumption among flats within buildings connected to district heating networks or central heating systems. In recent years, HCA data has also been utilized for building heating system analysis, fault detection, diagnosis, and optimization. However, certain inherent limitations of HCAs, such as data truncation and the sparse recording of data points over time, can hinder their direct application in analysis. This underscores the necessity of pre-processing HCA data prior to conducting meaningful analyses. This study aims to develop a methodology for recovering the decimal values of HCA data. By leveraging the continuity of HCA increments and the inverse relationship between external temperature changes and HCA increments, the problem is formulated as an optimization problem. Two case studies were conducted to validate this method. The first case study involved a radiator heating laboratory at the National Metrology Institute of Italy (INRIM), where 40 radiators of various types, geometries, and materials were tested. The lab replicates heating operations typical of real apartment buildings, utilizing specific control strategies and flexible hydraulic connections. The second case study focused on a residential building in Denmark, analyzing HCA data collected from 15 apartments over one month. In both case studies, we used different measures to collect HCA data with decimals as the reference. Results indicate that the proposed method significantly reduces errors and uncertainties associated with data truncation in both laboratory and real-world settings. On average, the root mean square error (RMSE) of the recovered HCA data compared to the reference value decreased by 76.9% and 60.4% when compared to the truncated data in the lab and real buildings, respectively. This demonstrates the

---

Q. Yang (✉) · M. Tunzi
DTU Construct, Technical University of Denmark, Lyngby, Denmark
e-mail: qyan@dtu.dk

Q. Yang · R. Salenbien
Flemish Institute for Technological Research (VITO NV), Mol, Belgium

EnergyVille, Genk, Belgium

F. Saba · M. Orio · M. Santiano · E. Audrito
National Metrology Institute of Italy (INRIM), Turin, Italy

© The Author(s) 2026
D. Vanhoudt (ed.), *Proceedings of the 19th International Symposium on District Heating and Cooling*, Lecture Notes in Networks and Systems 1700,
https://doi.org/10.1007/978-3-032-09844-3_15

method's effectiveness in enhancing the usability and reliability of HCA data over short time intervals.

Keywords Heat cost allocators · Data preprocessing · Radiators · Energy meter

1 Introduction

Heat cost allocators (HCAs) are indirect meters installed on radiators and are widely used across European countries [1]. These devices measure the relative heat output of each radiator within space heating (SH) systems, enabling fair distribution of heating costs. According to the Energy Efficiency Directive (EED) [2], all energy meters and submeters in buildings should support remote data access. Given that HCAs share similar properties with energy meters, they hold great potential for heating system analysis and optimization. Previous studies have demonstrated that HCA data can be utilized to identify apartments with high heat demand, determine system supply temperatures [3], and detect non-uniform heat distribution [4], as well as diagnose possible misuse or suboptimal operation within SH systems.

Despite their potential, HCAs were originally designed for billing over entire winter seasons rather than capturing dynamic heating behavior. HCAs accumulate the estimated heat output over time, and the operators collect the information based on a time length equal to the billing period, which is typically one month or even longer. According to the technical specifications, the device displays only integer values on its screen, and the company prefers to align the billing data with what users directly observe [5]. This practice is common among meter manufacturers. This introduces a problem: data is logged only with the integer part, which introduces significant uncertainties when looking at a shorter time interval compared to the billing time range, such as one or a few hours, and in particular for applications of the system optimization. To be more specific, the maximum uncertainty caused by truncation is +1 unit. In this study, the consumption of active radiators per day ranged from 4 to 20 units during the sampling period. Therefore, truncation has a substantial impact on the daily analysis, and this effect becomes even more significant for those low-consumption radiators and when shorter time intervals are considered.

To address this problem, this study aims to develop a methodology to reconstruct the missing decimal values in HCA data. By filling in these gaps, the methodology will enhance the precision of dynamic SH system analyses, ultimately improving the reliability of using HCA data for heating system optimization and supporting its future use. Some studies have attempted to infer and reconstruct the decimal parts of smart meter readings [6, 7]; however, to the best of the author's knowledge, there is currently no research focusing on the reconstruction of the decimal parts of HCA data.

2 Methodology

The methodology section will describe the counting basics of the HCA, the assumptions, and the methodology.

HCAs are counting the heat output from the radiators based on the integral of the temperature difference between the radiator surface and the indoor air. According to the European standard of HCAs EN834 [8], the HCA progression can be expressed by Eq. 1:

$$R = K_Q K_C K_T \int \left(\frac{T_{rad} - T_{in}}{60}\right)^n dt = K \int \left(\frac{T_{rad} - T_{in}}{60}\right)^n dt \qquad (1)$$

where:

R is the HCA progression, representing the accumulated value that reflects the energy output of the radiator.

K_Q is the rating factor for the thermal output of the radiator, representing the non-dimensional numerical value of the reference output of the radiator.

K_C: The rating factor for the thermal coupling of the sensors.

K_T: The rating factor for rooms with low design indoor temperatures.

K: The total rating factor, the product of the individual rating factors.

T_{rad}: The radiator surface temperature measurement.

T_{in}: The indoor air temperature.

n: The radiator component.

60: The average temperature at the design condition of 90/70/20 °C.

In this study, there is no different thermal coupling of the sensors, nor low temperature setpoints, hence K_C and K_T are treated with default values. As a result, the total rating factor is only determined by the characteristics of the radiators.

The working principle of radiators and the nature of the SH system derive two assumptions that support the development of the methodology.

1. The radiator heat output should be continuously changing from one time interval to another. This is based on the heavy thermal inertia of the hydronic system.
2. The radiator heat output should react to the outdoor temperature in a negative relationship. This is defined based on the fact that the SH system in apartment buildings normally follows a weather compensation control, in which the supply temperature is adjusted against the change of the outdoor temperature.

Hence, the optimization problem is defined as below.

$$\min_{\hat{y}} J(\hat{y}) = \sum_{i=2}^{n-1} (\Delta \hat{y}_{i+1} - \Delta \hat{y}_i)^2 + \lambda \sum_{i=2}^{n} \Delta \hat{y}_i \Delta T_i$$

$$s.t. y_i^{trunc} \le \hat{y} < y_i^{trunc} + 1 \qquad (2)$$

where:

$\hat{y}_i$: Recovered HCA progression at sampling point i

y_i^{trunc}: Observed HCA progression at sampling point i

T_i: Outdoor temperature at sampling point i

$\Delta \hat{y}_i = \hat{y}_i - \hat{y}_{i-1}$: Recovered HCA increment.

$\Delta T_i = T_i - T_{i-1}$: Outdoor temperature increment.

λ: Weighting coefficient between the two terms.

The first term reflects the continuity of the HCA increment change, and it penalizes abrupt changes in energy increments, promoting temporal smoothness. While the second term encourages a negative correlation between energy increments and outdoor temperature increments, ensuring consistency with the weather compensation control in the SH system. Especially, the initial guess of the decimal will start from 0.5 since this is the middle point of the interval.

3 Case Study

In this analysis, two case studies were selected for testing and verifying the methodology. The first one is a laboratory environment, the second one an existing residential building.

3.1 Italian Case—INRIM Lab

The first dataset originates from a laboratory environment at the National Metrology Institute of Italy (INRIM). This laboratory, known as the Heat Accounting Experimental Mock-Up, is specifically designed for the calibration and evaluation of HCAs [9].

The test facility features a central heating system powered by a gas boiler, connected to 40 radiators of various types, sizes, and materials. These radiators are distributed across four floors and linked through hydraulic loops, which can be easily configured for either vertical riser or horizontal connections. Each radiator is equipped with an HCA for data recording and a Direct Heat Meter (DHM) for the reference measurement of the radiator heat output, traceable to the International System of Units (SI). The lab also has a temperature sensor to record the outdoor temperature.

In February 2025, an experiment using all 40 radiators was conducted. For this experiment, only integer values are obtained through the billing system. To acquire

Table 1 Characteristics of radiators in the INRIM lab

Radiator type	Number	Capacity (W)	Original K	Modified K
Cast aluminum sectional 1	16	1467	1.807	10
Cast aluminum sectional 2	4	815	1.026	10
Four columns of cast iron 1	4	1427	1.923	10
Four columns of cast iron 2	2	713.5	0.997	10
Four columns of tubular steel 1	4	1482	1.859	10
Four columns of tubular steel 2	2	798	1.041	10
Heated towel rail	8	395	0.306	10

decimal values, we manually increased the HCA's rating factor. The specific details of this adjustment are presented in the Table 1.

After increasing the rating factor for all the radiators to the maximum value of 10, the progression of the HCA enlarged to 5.2–32.5 times, which is equivalent to adding the decimals of 0.2–0.03. For large radiators, the effect didn't strictly reach "adding one decimal", but in this study, the HCA data with decimals is only used as references rather than inputs; hence, this will not give a big effect on the final results.

3.2 Danish Case—Viborg Apartment Building

The second dataset originates from a multi-story residential building in Viborg, Denmark, constructed in the 1970s. We focused on a section including 15 apartments equipped with 84 radiators, each mounted with HCA. The data collection period spans January 2021. Typically, HCA data in real-world buildings is only available via the billing system and recorded as integers. However, for this project, we obtained high-resolution backend HCA data with decimal values directly from the HCA provider. Weather data for Viborg was sourced from the Danish Meteorological Institute (DMI).

4 Results and Discussions

The proposed method was applied to raw data from the INRIM laboratory. Results from radiator #4 are shown in Fig. 1, illustrating the HCA increments of the truncated data, the reference values with decimals, and the recovered data. The radiator was randomly selected for illustration purposes. As shown, the original truncated HCA increments fluctuate between integer values, while the recovered data closely follows the reference curve.

Similarly, Figs. 2 and 3 present results from the case study conducted in Viborg. In both cases, the recovered curves show a strong agreement with reference data.

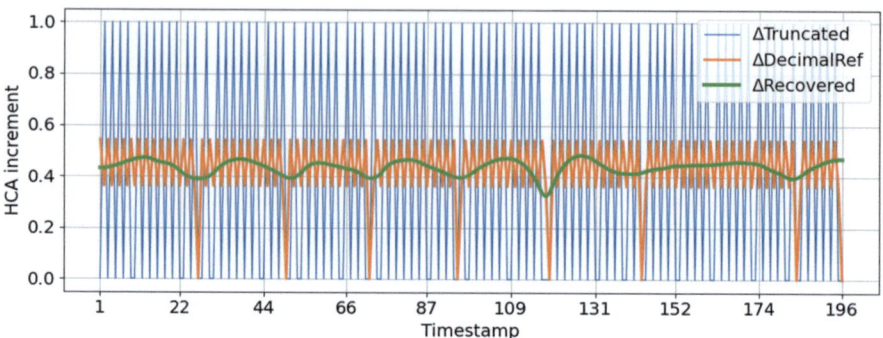

Fig. 1 HCA increment of INRIM radiator #4

Notably, Fig. 3 demonstrates that even when the reference HCA increments exhibit varying jump ranges (e.g., between 0–1 or 0–2), the recovered values still closely track the reference curve, highlighting the robustness of the proposed approach.

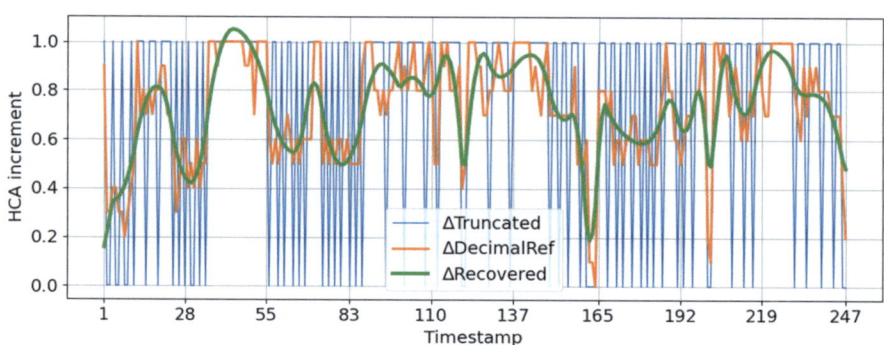

Fig. 2 HCA increment of Viborg radiator #1

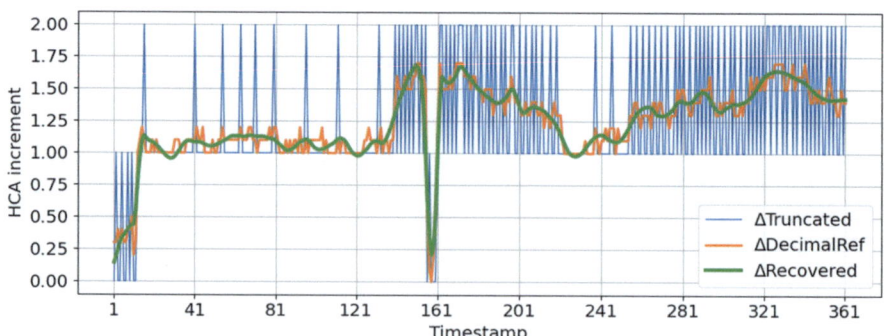

Fig. 3 HCA increment of Viborg radiator #4

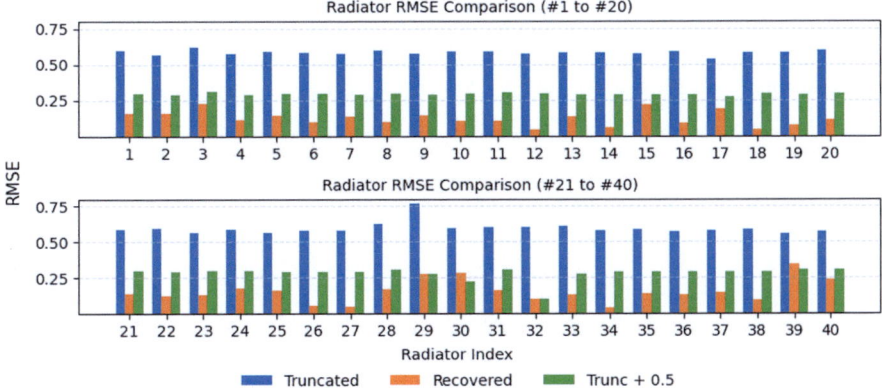

Fig. 4 RMSE between the reference true value and the truncated, recovered, and plus-0.5 guess data for all radiators in the INRIM lab

The data also highlights the difference between lab conditions and real buildings: the heat output from the lab radiator was much more stable compared to that of the radiators in the real building.

For an overall picture of the performance of the methodology across all radiators, Figs. 4 and 5 summarize the root mean square error (RMSE) of the truncated value, recovered value, and the truncated value plus 0.5 compared to the reference value, for both the INRIM and Viborg cases. The value 0.5 comes from the midpoint of the truncation interval and represents a simple guess to address the truncation issue.

In the INRIM lab, recovered data reduced the RMSE by 76.9% compared to the truncated data, while the plus 0.5 guess method reduced it by 51.6%.

The main difference between the lab and real building environments is that in the lab, we can operate all radiators, whereas in real buildings, some users may choose to use only part of the system and keep some radiators turned off. Among the 84 radiators, 18 were completely off, such as #8, #43, and #83. Excluding these radiators, the average RMSE reduction achieved by the recovery method is 60.4%, while the plus 0.5 method achieves a reduction of 44.3%. Specifically, the results for radiators #11 and #32 exhibited an unusual behavior: the truncated data showed a low RMSE, while the recovered data yielded a significantly higher RMSE. Upon examining the original data, it was observed that these radiators alternated between on and off states. Since the current method lacks the ability to detect the operational status of the radiators, it continued to compensate for gaps even during periods when the radiator was off. This inappropriate compensation led to amplified errors and, consequently, a higher RMSE. In future work, we plan to enhance the method by incorporating mechanisms to better identify the radiator's on/off status and thus avoid this issue.

The method demonstrated good RMSE performance in both laboratory and real-building settings, confirming its robustness across different conditions. Building on this, the improved reliability of short-interval HCA data can support more refined

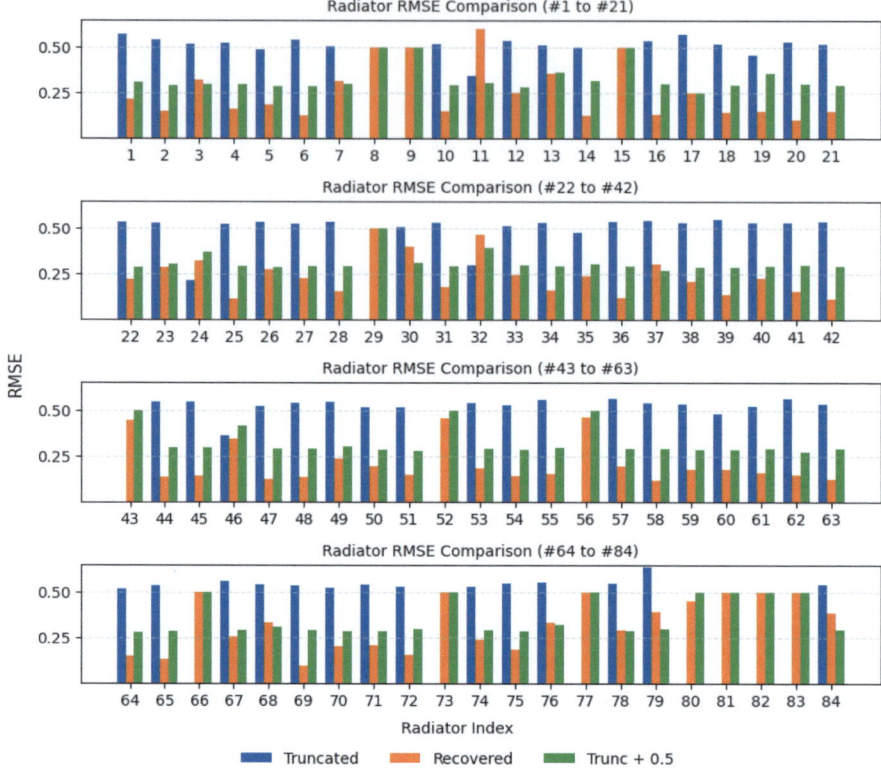

Fig. 5 RMSE between the reference true value and the truncated, recovered, and plus-0.5 guess data for all radiators in the Viborg building

analyses, such as identifying faulty radiators, detecting high-load apartments, and estimating the minimum required supply temperature for a given building. While HCAs are not designed for control purposes, the enhanced data quality can still contribute to radiator-level diagnostics and basic operational observations.

5 Future Work

An interesting direction for future research concerns the role of the weighting coefficient λ in the objective function. In this study, λ was used to balance two terms: temporal continuity and correlation with outdoor temperature. Preliminary analysis, as illustrated in Fig. 6 for one representative radiator, suggests that increasing λ generally leads to a decline in recovery performance. This might indicate that the simple correlation with outdoor temperature does not directly improve accuracy in the current setting.

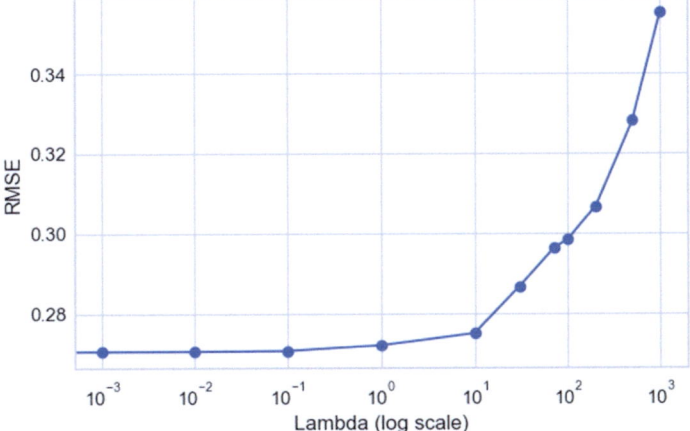

Fig. 6 RMSE between the recovered HCA increment and the true value under different weighting coefficients of Viborg radiator #1

One possible explanation is that the thermal inertia of the heating system introduces a time delay between the radiator's response and the outdoor temperature changes. If this delay is not accounted for, the correlation term may introduce noise rather than guidance. Exploring this potential time shift and more advanced correlation structures is beyond the scope of the current study, but presents a promising avenue for future investigation.

6 Conclusion

Although HCAs are primarily designed for cost allocation rather than energy metering, their data holds potential for future applications such as fault detection and hydronic system optimization, especially in contexts requiring high temporal resolution. This study proposes a method to recover the truncated decimal values in HCA progression. Under laboratory conditions, the method reduces the RMSE of HCA increments by 76.9%, and by 60.4% in real building scenarios. By significantly improving the accuracy of HCA data, the proposed method enhances its usability beyond billing, enabling more reliable high-resolution analyses. As such, it offers an effective solution to the truncation problem and supports the broader application of HCA data in future hydronic system assessments.

Acknowledgements We gratefully acknowledge the support and collaboration of Brunata and ZENNER International, and Viborg Varme.

References

1. Canale L, Dell'Isola M, Ficco G, Cholewa T, Siggelsten S, Balen I (2019) A comprehensive review on heat accounting and cost allocation in residential buildings in EU. Energy Build 202:109398. https://doi.org/10.1016/j.enbuild.2019.109398
2. The European parliament and the council of the European union, directive 2012/27/EU of the European parliament and of the council of 25 October 2012 on energy efficiency, amending Directives 2009/125/EC and 2010/30/EU and repealing Directives 2004/8/EC and 2006/32/EC (Text with EEA relevance). http://data.europa.eu/eli/dir/2012/27/oj, 2012
3. Tunzi M, Benakopoulos T, Yang Q, Svendsen S (2023) Demand side digitalisation: a methodology using heat cost allocators and energy meters to secure low-temperature operations in existing buildings connected to district heating networks. Energy 264:126272. https://doi.org/10.1016/j.energy.2022.126272
4. Yang Q, Salenbien R, Smith KM, Tunzi M (2024) Identifying untraced faults associated with high return temperatures from heating systems in buildings connected to district heating networks. Energy 309:133097. https://doi.org/10.1016/j.energy.2024.133097
5. Danfoss, Thermostatic sensors type RA 2000. Accessed Jun. 19, 2025. [Online]. Available: https://assets.danfoss.com/documents/107870/AI015186402271en-010601.pdf
6. Leiria D, Johra H, Marszal-Pomianowska A, Pomianowski MZ (2023) A methodology to estimate space heating and domestic hot water energy demand profile in residential buildings from low-resolution heat meter data. Energy 263:125705. https://doi.org/10.1016/j.energy.2022.125705
7. Schaffer M, Leiria D, Vera-Valdés JE, Marszal-Pomianowska A (2023) Increasing the accuracy of low-resolution commercial smart heat meter data and analysing its error. https://doi.org/10.35490/EC3.2023.208
8. EN 834:2013 Heat cost allocators for the determination of the consumption of room heating radiators—appliances with electrical energy supply (2013)
9. Marinari C, Saba F, Grisotto F, Orio M, Masoero M (2015) The INRiM thermo-hydraulic mock-up for thermal energy measurement devices: design, construction and metrological characterization. Energy Procedia 78:2304–2309. https://doi.org/10.1016/j.egypro.2015.11.378

Open Access This chapter is licensed under the terms of the Creative Commons Attribution 4.0 International License (http://creativecommons.org/licenses/by/4.0/), which permits use, sharing, adaptation, distribution and reproduction in any medium or format, as long as you give appropriate credit to the original author(s) and the source, provide a link to the Creative Commons license and indicate if changes were made.

The images or other third party material in this chapter are included in the chapter's Creative Commons license, unless indicated otherwise in a credit line to the material. If material is not included in the chapter's Creative Commons license and your intended use is not permitted by statutory regulation or exceeds the permitted use, you will need to obtain permission directly from the copyright holder.

Foam Density Distribution Analysis in Pre-insulated Pipes Using Non-destructive X-Ray Microscopy

Pakdad Langroudi and Ingo Weidlich

Abstract Pre-insulated bonded pipes used in district heating systems are engineered for a service life of at least 30 years. The longevity of these systems is primarily influenced by the properties of the polyurethane (PUR) foam insulation. While operational and installation parameters contribute to overall durability, the characterization of the foam layer is critical for service life prediction. According to EN 253, pipe manufacturers are required to maintain a minimum foam density of 55 kg/m^3. However, the foaming process leads to a non-uniform density distribution across the pipe cross-section—from the steel carrier pipe to the high-density polyethylene (HDPE) casing. Determining the foam density distribution function enables normalization of measurement data, which is essential for subsequent analysis using spectroscopic techniques such as Laser Induced Breakdown Spectroscopy (LIBS) and Fourier Transform Infrared (FT-IR) Spectroscopy. Traditional methods for density determination, including liquid displacement, and the cut-and-weigh method, are often labor-intensive and prone to inaccuracies. To address these limitations, a methodology has been developed that utilizes nanoscale X-ray microscopy and computer vision techniques to analyze foam density distribution.

Keywords Computer vision · Data normalization · Distribution function · Rigid polyurethane foam · X-Ray scan

1 Introduction

Pre-insulated bonded (PIB) pipes in district heating (DH) systems are broadly being used and hold the majority share of the network pipes [1], which are expected to function effectively for a minimum of 30 years. A critical factor influencing the PIBs

---

P. Langroudi · I. Weidlich (✉)
HafenCity Universität (HCU), Hamburg, Germany
e-mail: ingo.weidlich@hcu-hamburg.de

P. Langroudi
e-mail: pakdad.langroudi@hcu-hamburg.de

durability is the quality of the rigid polyurethane foam (RPF) layer. This foam serves a dual purpose: it provides thermal insulation and acts as a load-bearing element that resists shear loads from soil-pipe interaction. According to the EN 253 standard, the minimum required foam density is 55 kg/m^3; however, the density is not uniform across the cross-section of the pipe. This variation is attributed to the foaming process and the pressure gradient extending from the center toward the steel medium pipe and the polyethylene casing.

Traditional methods for analyzing foam density distribution—such as Liquid Displacement, Gas Pycnometry, and the Cut and Weigh Method—often lack precision due to sample preparation challenges and the porous nature of the foam. Density distribution is key to understanding foam degradation, as the operating temperature profile varies from the medium pipe (highest temperature) to the outer casing (lowest). Accurate profiling of foam density is also essential for ensuring the reliability of analytical techniques such as Laser-Induced Breakdown Spectroscopy (LIBS) and Fourier Transform Infrared (FT-IR) Spectroscopy. For example, LIBS is used to study elemental changes, but translating spectral intensities into quantities requires knowledge of local density variations. Since the foam does not exhibit uniform intensity, understanding its density profile is critical. The same holds true for FT-IR and other spectroscopic methods, where absorption and reflection are influenced by local foam density. This paper introduces the non-destructive X-ray Microscopy (XRM) to precisely map the foam density distribution. The use of computed tomography (CT) for studying RPF is a well-established method employed by various researchers for a range of purposes, including analyses of cell structure performance, molecular structure and mechanical properties, and morphological assessments [2–5]. However, none of these studies specifically address density distribution, and 3D X-ray CT methods are not well-suited for the analysis of a large number of samples due to their time-consuming nature and significant hardware requirements. In this paper, four different methods for calculating density distribution are evaluated, with particular emphasis on both the time efficiency and accuracy of each technique, as discussed in the following sections.

2 Material and Methodology

2.1 Specimen Preparation

The specimens used in this study were derived from pre-insulated bonded pipes, commonly utilized in DH systems with the blowing agent cyclopentane. The samples were sectioned longitudinally to expose the cross-section for X-ray Microscopy (XRM) analysis. For the purposes of this study, the samples were prepared in a square cross section, and the height of the samples were equal to the height of the insulation thickness of the pipe (Fig. 2a).

2.2 X-ray Microscopy (XRM) Setup

The XRM was performed using a Nanoscale X-ray Microscope capable of capturing detailed true spatial resolution below 500 nm. This non-destructive imaging technique **was selected for its high resolution and ability to** penetrate the foam material without causing any physical alterations. The setup is equipped with a Xray with energy up to 160 keV and source power up to 16 W and large-format 16 MP CMOS detectors.

2.3 Data Acquisition and Analysis

Three data acquisition techniques have been used for capturing images of the samples. The aim of trying various acquisitions is to find the best compromise by lowering the acquisition time and maximizing the accuracy in measurements. Therefore, a 3D measurement ran with the length of the whole specimen, a 2D single-shot Xray with the length of the whole specimen, and a high-resolution stacked shots were performed at different segments of the specimen on Z axis (Table 1). Further details about the acquisition techniques are provided by He et al. [2].

The final data generated on all techniques were 16-bit grayscale tiff files that have been used for analysis or 3D reconstruction with computer vision algorithms. The density along the cross-section of the specimens was calculated by analyzing the grayscale intensity variations in the XRM images, which correspond to the material's density.

2.4 Data Normalization

To ensure consistency and comparability of the results, the acquired data is required to be normalized against a reference standard. This reference standard does not exist for RPF and as the aim of this research is to provide density information of the samples, each sample has been normalized against its own total density to eliminate the potential density related discrepancies such as volume, cell size, wall thickness, etc. This normalization process accounted for variations in the XRM system and provided a consistent baseline for comparing different specimens.

Table 1 Details of the three acquisition techniques utilized in this research

Designation	Description	Magnification (μm)	Acquisition time (sec)
2D	Single shot	15–30	~28
2D stack	Multiple shots across Z axis	4	~140
3D	Rotational imageing accoss Z axis	15–30	~21,600

The N is the Normalization Factor, which is given by formula (1).

$$N = (\rho)/(I) \tag{1}$$

where:

N = Normalization Factor.

ρ = Density of the specimen (g/cm³).

I = Total Inverted Grayscale Value of the Object.

The I in the formula is the sum of the pixel values of the image after inversion. The inversion is applied to correlate higher densities with higher grayscale values.

3 Results

3.1 Foam Density Distribution Profile

The XRM-based methodology provided a detailed density distribution profile across the foam layer with precision intervals relative to the acquisition magnification. The data revealed variations in density, particularly at the interfaces with the steel medium pipe and the high-density polyethylene casing (Fig. 1).

The results obtained using XRM were asserted with the conventional Cut and Weigh Method. The XRM approach demonstrated superior resolution, capturing density porous variations due to bubbling effect in foaming process and detailing of the spatial information that was not detectable using traditional techniques.

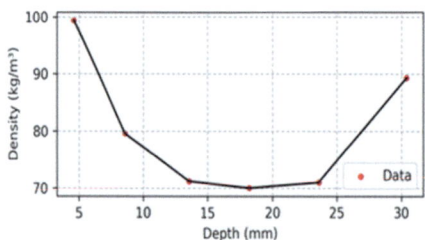
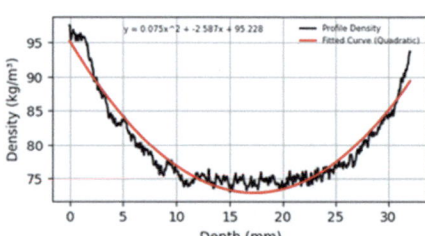

Fig. 1 Density Distribution Profile of the Specimen: The graph illustrates a distinct density gradient, with higher densities concentrated along the edges and a noticeable reduction in density toward the center of the profile. Cut and weight method (left) and XRM-based method (right)

3.2 Data Acquisition Techniques and Model Quality

X-ray microtomography is a technique for obtaining spatial information about samples. The 3D scanning method involves rotating the sample at specified intervals and reconstructing its structure based on the acquired data. Figure 2e shows a reconstructed sample along with examples of cross-sectional slices, scanned at a resolution of 25 µm.

By calculating the mean grayscale value of each slice, it is possible to generate a distribution profile of the specimen. However, the grayscale intensity units must be converted into mass for interpretation. To achieve this, each specimen was weighed, and the total mass was set equal to the sum of the grayscale values across the entire specimen. This calibration process translates intensity values into mass. Additionally, specimens were physically sectioned, weighed, and their dimensions measured to validate the predicted values against actual measurements. This procedure was repeated across all methods and specimens, with a minimum of three specimens analyzed per method. Although the 3D scanning approach provides highly detailed information about the specimen, it is computationally and hardware-intensive. To improve processing speed, alternative methods such as single-shot 2D imaging and 2D stacking were explored (Fig. 2d). In the 2D stacking technique, the specimen is imaged at a higher magnification (4 µm in this study), and the resulting images are stacked to form a composite representation of the sample. This method enhances the level of detail captured.

The scanned data were analyzed using linear regression to evaluate the accuracy of the different methods. While the fit quality was not optimal for the 2D scanning methods, the results prompted further investigation. For the single-shot 2D method, the regression curve was nearly parallel to the ideal fit but exhibited a consistent shift, likely due to sensor calibration errors. These errors may be intrinsic to the setup, which is optimized for 3D scanning, and are potentially mitigated during 3D reconstruction.

To address these issues, new samples were prepared, scanned, sectioned, and weighed. Despite implementing flat-field correction functions for the sensor, the quality of the regression fit did not improve substantially. However, the data points

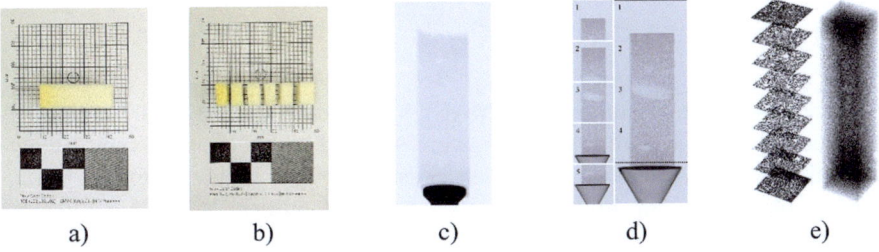

Fig. 2 Sample prepared for the study (**a**), the specimen after cut and weight (**b**), single shot 2D scan (**c**), stack shot 2D scan (**d**), 3D scan (**e**)

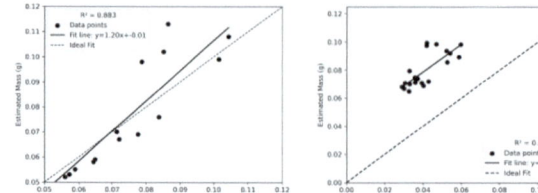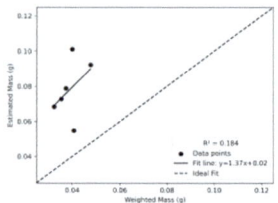

Fig. 3 Linear regression analysis of sample weights: The x-axis represents the weighted masses of the samples, while the y-axis shows the estimated weights derived from the model. The plot includes the linear fit line, the corresponding R2 value, and the model. 3D scan (left), 2D scan (mid), 2D-Stack (right)

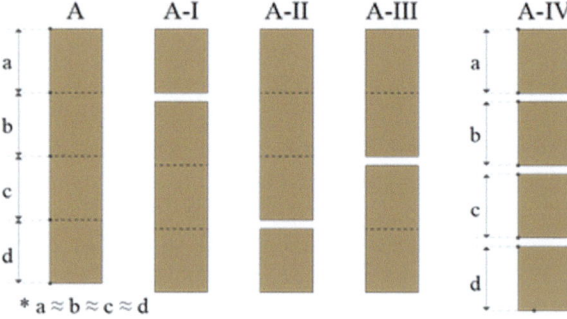

Fig. 4 Data augmentation technique to improve the statistical points in the dataset

aligned more closely along the ideal fit trajectory (Fig. 3). Subsequent analysis employed a data augmentation technique. Each specimen was sectioned into four segments, and their dimensions and weights were used to create an expanded dataset of four new specimens per sample (Fig. 4).

Using the augmented dataset significantly improved the quality of the regression fit, as demonstrated by the updated performance metrics (Fig. 5). This advancement underscores the potential of data augmentation to enhance the reliability of XRM analysis while addressing limitations in sensor calibration and data quality.

4 Discussion

The metrics—Mean Absolute Error (MAE), Mean Squared Error (MSE), and Mean Absolute Percentage Error (MAPE)—were selected to comprehensively evaluate prediction accuracy. Together, these metrics assess both absolute and relative prediction performance while highlighting the impact of outliers. All the metrics are summarized in Table 2.

The evaluation metrics provide a quantitative assessment of the predictive performance of the model. The **Mean Absolute Error (MAE)** of 0.0387 indicates that, on

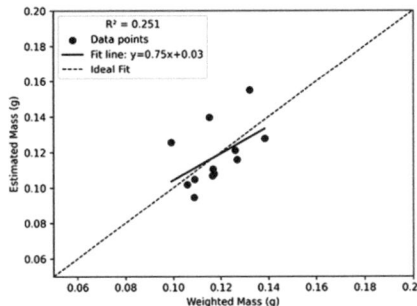 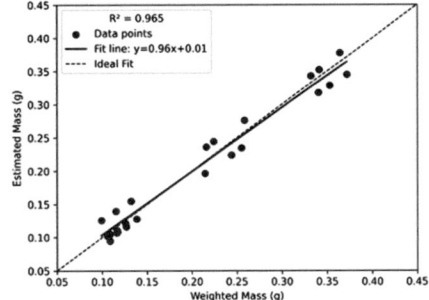

Fig. 5 Regression analysis of the 2D scanned samples after flat field calibration and mount adjustment (left). Data augmentation technique to improve the statistical points in the dataset. Regression analysis after data augmentation (right)

Table 2 Comparative performance of 2D, 3D, and Augmentation Methods based on error metrics

Method/ Metric	MAE	MSE	MAPE (%)	Observations
2D Method	0.0387	0.00155	49.23	Weakest performance, largest errors
3D Method	0.00748	0.000098	11.25	Significant improvement, reduced errors
Augmentation method	0.0162	0.00031	9.70	Best proportional accuracy

average, the predicted mass values deviate from the actual mass values by approximately 0.0387 units. The **Mean Squared Error (MSE)** of 0.00155 suggests that the model's errors are relatively small but highlights that larger deviations are penalized more heavily due to the squaring of errors. However, the **Mean Absolute Percentage Error (MAPE)** of 49.23% indicates that the model's predictions deviate, on average, by nearly half the actual values when expressed as a percentage. This relatively high MAPE suggests that the model may not be sufficiently accurate for applications where precise mass predictions are critical. These results point to the need for further refinement of the model, such as through feature engineering, hyperparameter tuning, or using a different modeling approach, to improve its predictive accuracy.

The evaluation metrics for the 3D method demonstrate a substantial improvement in predictive accuracy compared to the 2D method. The **MAE** for the 3D method is 0.00748, which is significantly lower than the MAE of 0.0387 obtained using the 2D method, indicating that the 3D method produces predictions that are much closer to the actual mass values. Similarly, the MSE of 0.000098 for the 3D method is markedly smaller than the 2D method's MSE of 0.00155, showing a substantial reduction in the magnitude of larger errors.

Furthermore, the **MAPE** of 11.25% for the 3D method represents a notable improvement over the 49.23% MAPE from the 2D method. This considerable decrease in MAPE indicates that the 3D method offers more reliable and consistent predictions. The 3D method's improved accuracy likely stems from its ability to capture foam microstructure, which strongly influences density and mass, as outlined by [6] in their study of cellular solids.

Overall, these results suggest that the 3D method is a more effective approach for predicting mass, likely due to its ability to capture additional spatial or structural information that the 2D method cannot.

The results for the 2D single shot method, which incorporates data augmentation and flatfield correction, demonstrate further refinement in predictive accuracy compared to both the 2D and 3D methods. The **MAE** of 0.0162 indicates a moderate increase in error compared to the 3D method (0.00748) but is still a marked improvement over the 2D method (0.0387). Similarly, the **MSE** of 0.00031 is higher than the 3D method's 0.000098 but remains significantly lower than the 2D method's 0.00155, showing that the final round method maintains a low error magnitude.

Most notably, the **MAPE** of 9.70% is the lowest among the three methods, reflecting the best performance in terms of proportional accuracy. This improvement suggests that the final round method effectively aligns predictions with actual values, particularly in scenarios where relative accuracy is crucial.

In comparison:

- The **2D method** had the weakest performance, with the largest errors across all metrics.
- The **3D method** showed significant improvement, particularly in reducing absolute and squared errors.
- The **Augmentation method** further enhanced proportional accuracy (lowest MAPE) while maintaining relatively low absolute and squared errors.

4.1 Future Work

Future research should focus on further refining the computer vision algorithms used in this study to enhance their accuracy and reduce processing time. Additionally, exploring the application of XRM in other materials and industries could provide valuable insights into the broader applicability of this methodology.

5 Conclusion

This study presents a non-destructive methodology for profiling the density of polyurethane foam in pre-insulated bonded pipes using Nanoscale X-ray Microscopy (XRM). The proposed method offers significant improvements in accuracy and efficiency. The results obtained using XRM were compared with those from conventional methods. The XRM approach demonstrated superior accuracy and precision, capturing density variations that were not detectable using traditional techniques. Additionally, the non-destructive nature of the XRM method allowed for repeat analyses on the same sample, further enhancing reliability.

Acknowledgements The authors acknowledge the financial support by the Federal Ministry for Economic Affairs and Climate Action of Germany in the project SAM-FW (project number 03EN3078B).

Author Contribution Pakdad Langroudi was responsible for the study design and conception, data collection and analysis and chiefly responsible for preparing the manuscript. Ingo Weidlich provided review, editing and scientific supervision of the study.

References

1. Langroudi P (2022) Predictive maintenance: ageing and lifetime modeling of pre-insulated district heating pipes. unpublished PhD dissertation, HafenCity Universität Hamburg (HCU)
2. He Y, Qiu D, Yu Z (2021) Multiscale investigation on molecular structure and mechanical properties of thermal-treated rigid polyurethane foam under high temperature. J Appl Polymer Sci 138(44). https://doi.org/10.1002/app.51302
3. Nistor A, Toulec M, Zubov A, Kosek J (2016) Tomographic reconstruction and morphological analysis of rigid polyurethane foams. Macromolecular Symposia 360(1):87–95. https://doi.org/10.1002/masy.201500113
4. Morankar S, Mort R, Curtzwiler G, Vorst K, Jiang S, Chawla N (2024) Structural features of biobased composite foams revealed by X-ray tomography. RSC Adv 2024 4(27):19528–38. https://doi.org/10.1039/d4ra02461c][PMID: 38895520
5. Fu Y, Qiu C, Ni L, et al (2024) Cell structure control and performance of rigid polyurethane foam with lightweight, good mechanical, thermal insulation and sound insulation. Constr Build Mater 447:138068. https://doi.org/10.1016/j.conbuildmat.2024.138068
6. Gibson LJ, Ashby MF (1997) Cellular solids: structure and properties, 2nd edn. Cambridge University Press, Cambridge

Open Access This chapter is licensed under the terms of the Creative Commons Attribution 4.0 International License (http://creativecommons.org/licenses/by/4.0/), which permits use, sharing, adaptation, distribution and reproduction in any medium or format, as long as you give appropriate credit to the original author(s) and the source, provide a link to the Creative Commons license and indicate if changes were made.

The images or other third party material in this chapter are included in the chapter's Creative Commons license, unless indicated otherwise in a credit line to the material. If material is not included in the chapter's Creative Commons license and your intended use is not permitted by statutory regulation or exceeds the permitted use, you will need to obtain permission directly from the copyright holder.

A 5th Generation District Heating and Cooling Network (5GDHC) Driven by Shallow Geothermal, Economic Analysis and Geohydrological Modelling for Comparison with an Individual System

Hans Hoes and Jente Pauwels

Abstract The success of the technology concerning shallow geothermal installations creates a great density where systems will increasingly affect each other negatively. Especially in an urban context, this becomes an inextricable tangle. On top of that, different types of systems are active, both smaller closed and larger open systems are applied according to project size and specific needs. Based on a concrete example case, it is shown that it makes sense to go for a collective approach, in combination with a 5th GDHC. The impact is investigated by subsurface modelling and additional economic analysis. The hydrogeological modelling is performed using FeFlow 10.0, a 3D finite-element software. Both open loop systems (ATES, Aquifer Thermal Energy Storage) and closed loop systems (BTES, Borehole Thermal Energy Storage) are modelled. The thermal and hydraulic impact of the different systems on each other and the environment is investigated. The results show that when shallow geothermal systems are operating close to each other, there is a risk of mutual impact. Open loop systems can have both thermal and hydraulic impact on nearby systems, while closed loop systems can only thermally impact their surroundings significantly. Moreover, the economic analysis shows that the project is more cost-efficient by using a combined geothermal system. The use of a fifth-generation network with uninsulated pipes optimally contributes to favourable economics.

Keywords Geothermal · ATES · BTES · Economic analysis

H. Hoes (✉) · J. Pauwels
Terra Energy, Geel, Belgium
e-mail: hoes@terra-energy.be

© The Author(s) 2026
D. Vanhoudt (ed.), *Proceedings of the 19th International Symposium on District Heating and Cooling*, Lecture Notes in Networks and Systems 1700,
https://doi.org/10.1007/978-3-032-09844-3_17

1 Introduction

Both negative and positive interactions can occur due to the increasing density of shallow geothermal systems. Sub-optimal implantation of the wells of these geothermal systems will lead to reduced efficiency and poor profitability. In addition, this is detrimental to the pursuit of further sustainability of the building stock as it will quickly lead to saturation of the available space. Therefore, a study was set-up to investigate the economics and efficiency between a 5th generation district heating and cooling network (Scenario A) and individual geothermal systems (Scenario B). A case study was selected based on the presence of favourable conditions for an ATES (Aquifer Thermal Energy Storage) system and the availability of different services on a small area. In Scenario A, the different service are coupled in one large ATES system, while in Scenario B the different services are coupled in their own individual ATES or BTES (Borehole Thermal Energy Storage) systems. An overview of the available services in the case-study is shown in Fig. 1. The case-study consists of a sports park, a school with a sports hall, an administration building, a guest house, a meeting centre, service flats, social housing, a residential care centre and a swimming pool.

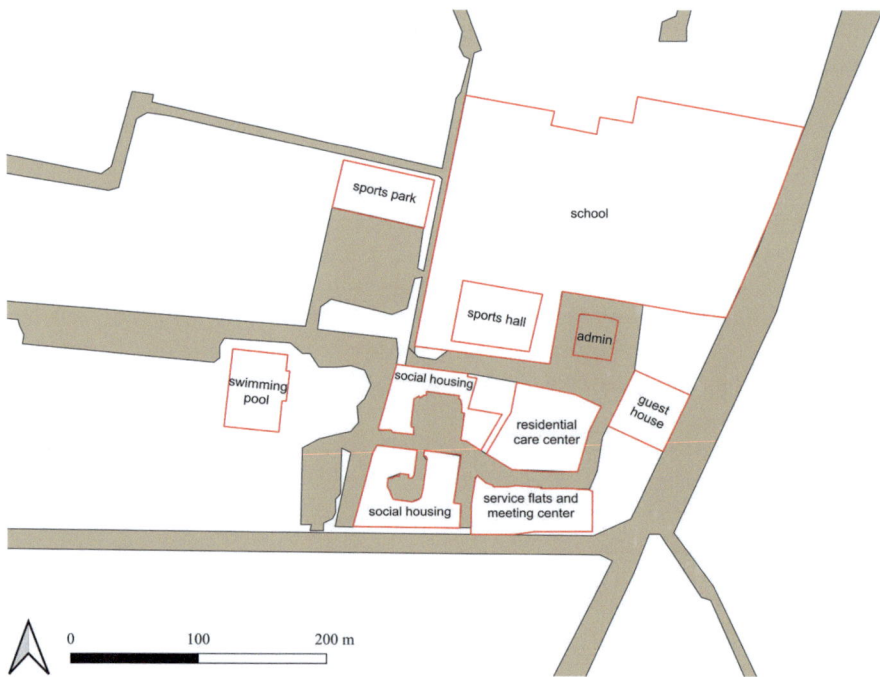

Fig. 1 Overview of the different services which could be connected to a 5th generation shallow geothermal system

2 Methodology

2.1 Economic Analysis

The development of geothermal energy projects often grows organically as needs arise. This leads to buildings being addressed one by one, with a clear shift in time. Often new buildings are installed, which may be given an immediate renewable energy supply. As a result, you get a gradual build-up of individual geothermal projects, each designed to suit individual needs. A collective approach can bring synergy benefits, especially if ATES systems with a large capacity can be developed, as is the case in this project. The design of such an approach is not straightforward; a vision must be developed whereby all parties involved enjoy the benefits. A competent operator takes care of the various customers by keeping the installation functioning optimally.

To make an economic analysis possible, a standard installation (called a reference) should be defined as the point of comparison. An individual air-to-water heat pump is considered as the standard reference. This is compared with the innovative solution with collective geothermal installation (scenario A). In this scenario, the geothermal energy (open system, type ATES) is brought to the individual consumers through a source network (5th generation heat grid). Each building has a substation with its own water-to-water heat pump and a passive cooling exchanger (if needed). In addition, the comparison is made with an individual geothermal system (scenario B). In this scenario, each application gets its own individual geothermal solution, the equipment in the technical rooms is similar to the situation in scenario A.

2.2 Subsurface Modelling

The software which is used for modelling the case study is FeFlow 10.0. It is a 3D finite-element software which simultaneously solves the mass, flow and heat transport equations [1]. A finite-element mesh is used to calculate the different parameters such as hydraulic head, temperature, flow… at cm-scale. The subsurface is build up out of (hydro)geological and topographical information which is implemented by the user. The used thermal and hydraulic parameters which represent the different geological layers are summarized in Table 1.

The model has a total depth of 175 m, until the bottom of the Miocene Aquifer system. In Scenario A, one large ATES system will be installed, consisting of four open-loop systems (a total of eight wells). In Scenario B, the different services will have their own ATES or BTES system. A summary of the different systems, including the annual energy demand, is summarized in Table 2. All ATES systems have filter settings in the Miocene Aquifer system. All BTES systems have a depth of 150 m below ground level.

Table 1 Summary of the hydraulic and thermal properties of the different layers of the model

Aquifer system	$K_{xx} = K_{yy}$ (m/day)	K_{zz} (m/day)	Specific storage (m^{-1})	Heat capacity ρ (MJ/m^3/K)	Heat conductivity λ (W/mK)
Quaternary	1.0	0.1	0.02	2.5	2.4
Clay-sand complex of the Campine	0.2	0.0091	3.69×10^{-5}	2.5	2.4
Pleistocene and Pliocene aquifer system	9.9	0.5	1.85×10^{-5}	3.2	1.6
Miocene aquifer system	10	0.2	1.16×10^{-5}	2.5	2.4

Table 2 Summary of the different geothermal systems used in Scenario A and B

Service	Scenario A		Scenario B		Annual heat demand (MWh)	Annual cool demand (MWh)
	System	Number of wells	System	Number of wells		
School, sports hall & park	Shared ATES system	8	ATES A	4	2013	230
Swimming pool			ATES B	2	1161	175
Guest house			ATES C	2	101	47
Residential care center			ATES D	2	811	367
Service flats & meeting center			ATES E	2	301	65
Social housing			BTES A	25	205	50
Administration building			BTES B	8	51	23

To simulate the thermal load on each of the ATES systems, pumping profiles (in m^3/chosen time step) are assigned to each well. Two wells are coupled to design an open loop system. To indicate the temperature change, a ΔT file is additionally assigned to each open loop system. To simulate the thermal load on each of the BTES systems, power profiles (in kW) are assigned to each system.

The initial groundwater flow is calculated based on the average groundwater level at t_0 at the boundaries of the model and the permeabilities of the different aquifer systems. The initial temperature distribution in the model is calculated based on the geothermal gradient ($\approx$2.9 °C/100 m-bgl). The simulations are run for 20 years to investigate the long-term consequences.

3 Results and Discussion

3.1 Thermal and Hydraulic Impact

Scenario A. The shared ATES system of scenario A shows a relatively small thermal and hydraulic impact on the environment (see Fig. 2). A separate cold and warm plume are formed in the aquifer. In Scenario A, the amount of wells is rather limited (eight in total), which results in relatively small cumulative hydraulic fluctuations. The maximum water level changes in the wells do not exceed ± 3,0 m. These changes are only temporary as when flowrate is lowered, the water level restores itself.

Scenario B. The thermal and hydraulic impact of the different systems is rather small. However, due to a greater density of individual systems, there is a chance on negative interference between all systems (Fig. 3). The warm wells of ATES systems B, D and E are located in between two cold plumes. The chance of thermal interference over time is large. The individual BTES boreholes of the social housing are located in the warm plume generated by ATES systems D and E. BTES system B from the administration building is located in the cold plume generated by ATES systems C, D and E. Due to the presence of the ATES systems nearby BTES systems, the performance of the BTES systems can significantly decrease. Figure 4 shows the fluid temperature out of a single BTES borehole with (blue line) and without (red line) the presence of a cold ATES well. A downward trend is observed in the situation where a cold well is present nearby the BTES borehole. This will decrease the overall efficiency of the BTES system as they are mostly used for heating purposes.

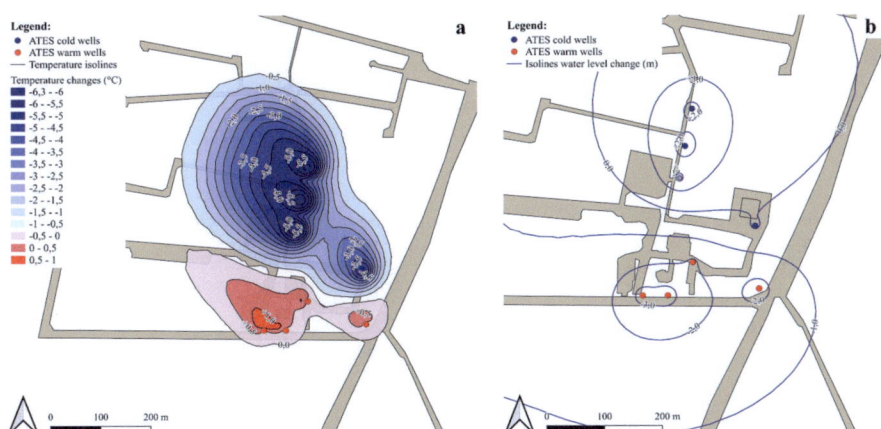

Fig. 2 Scenario A during winter period: **a** temperature changes (in °C) after 20 years of continuous operation and **b** hydraulic fluctuations (in m) during peak performance

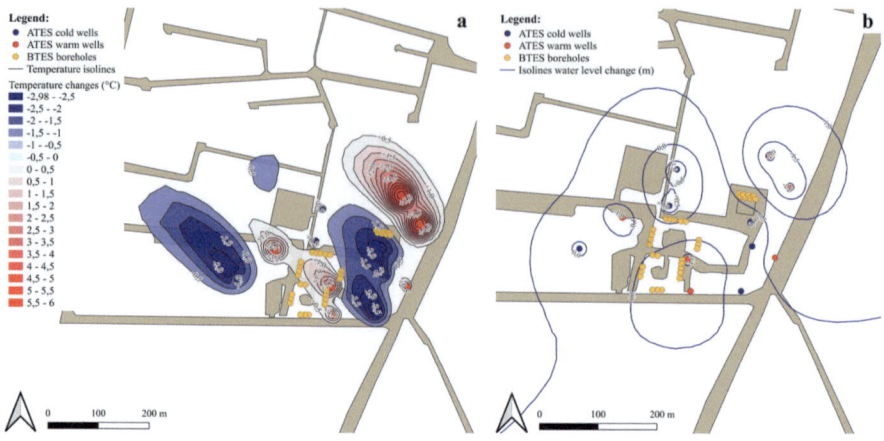

Fig. 3 Scenario B during winter period: **a** temperature changes (in °C) after 20 years of continuous operation and **b** hydraulic fluctuations (in m) during peak performance

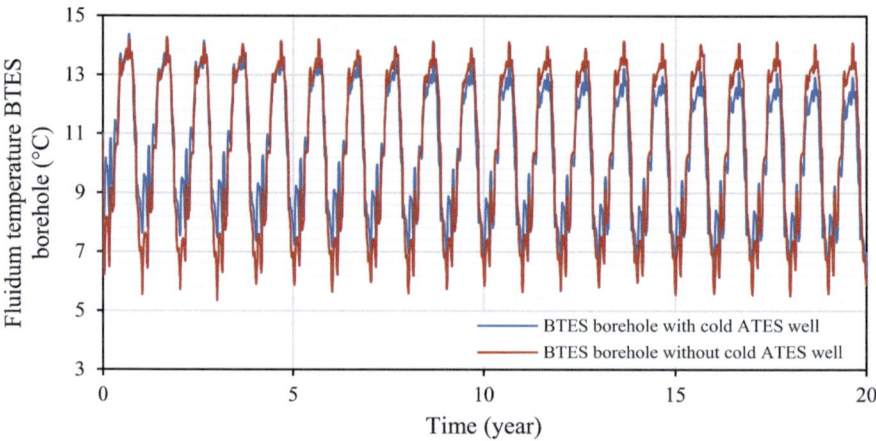

Fig. 4 Output fluid temperature of a BTES borehole with and without the close presence of a cold ATES well

3.2 Economic Analysis

Table 3 summarizes the investment cost, annual operating cost and the simple payback period of both scenarios (A and B) compared to the reference plant. The investment includes the cost of the heat pumps (air/water or water/water), the equipment of the technical rooms (piping, pumps, buffers, control system, expansion vessels, fittings and accessories), the investment of the individual or collective geothermal installation and the required source-side heat network (only in scenario A).

Table 3 Summary of costs and benefits

	Unit	Reference	Scenario A	Scenario B
Investment	EURO	€ 2.699.000	€ 4.353.500	€ 4.837.000
Extra investment compared to reference	EURO		€ 1.654.500	€ 2.138.000
Annual operating cost	EURO/year	€ 388.492	€ 211.618	€ 223.860
Savings on operating cost (compared to ref.)	EURO/year		€ 176.874	€ 164.632
Simple payback time	Year		9,4	13,0

This shows an additional cost on the investment of 1.6 M€ (scenario A) and 2.1 M€ (scenario B) compared to the classic reference. This is offset by annual savings on operating costs of 176.8 k€ (Scenario A) and 164.6 k€ (Scenario B). The operating cost includes all consumption and maintenance costs. The consumption concerns the electrical consumption of source and transfer pumps as well as the compressors of the heat pumps. The maintenance cost includes the annual maintenance actions of the heat pumps and the possible cost for annual monitoring of the geothermal installations (preventive monitoring, permit obligations such as environmental inspection reporting and water analyses). The simple payback period is 9.4 years (scenario A) and 13 years (scenario B) compared to the reference.

The Life Cycle Cost (LCC) is an evaluation process that can analyse the economic performance of a system, plant or building over its entire lifetime. LCC is a technique that compares the initial investments with the operational gains of installations or buildings. In this calculation, the investment costs, energy costs, maintenance costs and reinvestments of both plants (reference and geothermal plants) are used as inputs to the LCC. No account was taken of the residual value of the installations and the disposal cost as well as the costs for the distribution and delivery of the heat and cold in the building (pipe network, air conditioning ceilings, etc.). These installations are assumed the same in both analyses so no difference can be made here. The calculation of the LCC starts in the year 2025 (year of investment) and ends in 2055 (30 years). Basically, energy and maintenance costs are updated by adding annual inflation and an increase in energy prices with the energy factor. The sum of energy and maintenance costs is called the operating cost. The discounted energy and maintenance costs are added over its lifetime; this gives the total current cost of operating costs. The reinvestments are updated by adding inflation, this gives the total current cost for the reinvestments. The total current cost is the sum of the net present values of capital expenditures, operating costs and reinvestments and indicates what we need to pay all energy supply costs in the future. The lower the total current cost, the less money needs to be spent to keep the plant operational over its lifetime (Fig. 5).

The instability of energy prices necessitates an examination of the impact on the LCC for this project. When comparing systems, it is important to conduct a sensitivity analysis of the LCC input data. Figure 6 shows total net present value under different assumptions for the reference and geothermal systems over the lifetime.

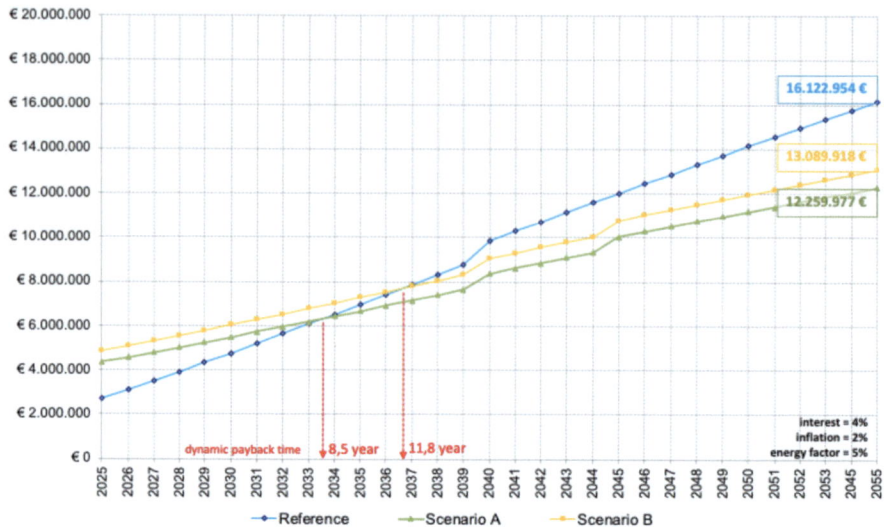

Fig. 5 Total actual cost of geothermal installations compared to the reference

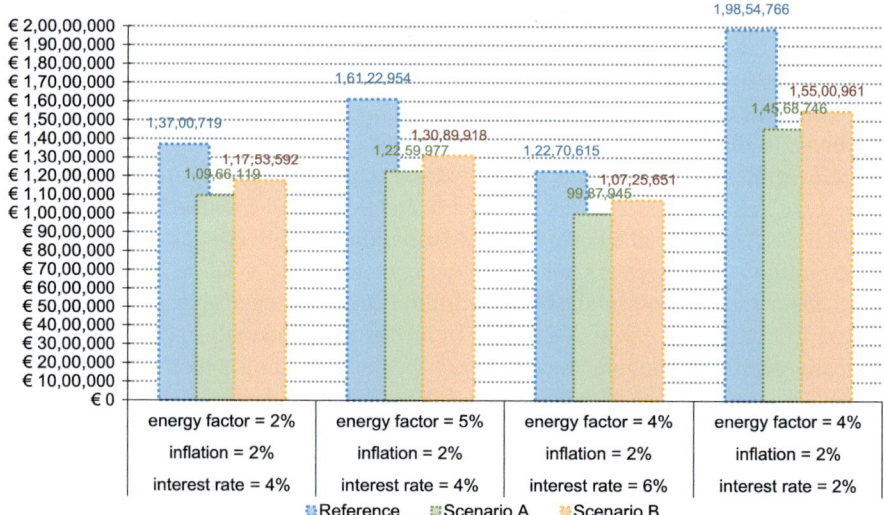

Fig. 6 Net present value on 30 years

Scenario A shows that both investment and operation are more advantageous than the individual solution (scenario B). In ATES, there are large synergy benefits to be gained by maximising the capacity of the well doublets. This is because the investment does not increase linearly with capacity. This more than offsets the construction of a heat network. The construction cost of the heat network (which is actually a source network) is much cheaper than a conventional heat network because the pipes do not need to be insulated. Furthermore, the source network operates as a large storage tank whereby buildings exchange heat and cold to each other.

4 Conclusions

As a result of the increasing density of systems positive and/or negative interactions between systems can occur. When negative interaction occurs, it can significantly decrease the efficiency of the impacted system. It is designated to look at the possibility of larger collective systems as a 5th generation district heating/cooling network. This way, different geothermal systems can be better coordinated to avoid conflicts in the future. The economic analysis also shows favourable conditions when a 5GDHC network is installed compared to individual systems. The additional cost of investment is higher when individual systems are installed (scenario B) compared to the collective system (scenario A). After evaluating the Life Cycle Cost of both scenarios, a dynamic payback time is calculated of 8,5 years for scenario A and 11,8 years for scenario B.

Reference

1. Diersch H-JG (2014) FEFLOW finite element modeling of flow, mass and heat transport in porous and fractured media. Springer Berlin, Heidelberg

Open Access This chapter is licensed under the terms of the Creative Commons Attribution 4.0 International License (http://creativecommons.org/licenses/by/4.0/), which permits use, sharing, adaptation, distribution and reproduction in any medium or format, as long as you give appropriate credit to the original author(s) and the source, provide a link to the Creative Commons license and indicate if changes were made.

The images or other third party material in this chapter are included in the chapter's Creative Commons license, unless indicated otherwise in a credit line to the material. If material is not included in the chapter's Creative Commons license and your intended use is not permitted by statutory regulation or exceeds the permitted use, you will need to obtain permission directly from the copyright holder.

Improved District Heating Return Temperatures by Cascading Concepts

Jan Eric Thorsen⑩, Oddgeir Gudmundsson⑩, Michele Tunzi⑩, and Marek Brand

Abstract As the focus on the performance of district heating (DH) systems intensifies, this study explores three cascaded substation concepts to assess their potential for reducing the DH return temperature at the building level substation. A lower DH return temperature is crucial for lowering the DH flow temperature to optimal levels, thereby enhancing system efficiency, which is a key feature of 4th generation DH (4GDH). Within the ARV project (https://greendeal-arv.eu), DH substation concepts have been evaluated, including parallel, two-stage, aftercooling, and midcooling configurations, with the parallel concept serving as the baseline for comparison. The analysis, based on annual simulations, covers generalized parameter combinations to demonstrate the potential for DH return temperature reduction across the different substation concepts in comparison to the baseline. Additionally, the impact of various climate profiles is explored, represented by the locations of Copenhagen, Helsinki, Paris, and Rome. Field data from a two-year test of the aftercooling concept validates the analysis results. The aftercooling and midcooling concepts have a significant reduction potential in annual DH return temperatures by 3 to 9,5 °C for 4GDH operations, compared to the baseline system.

Keywords District heating · 4th Generation district heating · Low-temperature district heating · District heating return temperature · District heating substation

---

J. E. Thorsen (✉) · O. Gudmundsson · M. Brand
DCS Solutions & Technology, Danfoss A/S Nordborgvej 81, 6430 Nordborg, Denmark
e-mail: jet@danfoss.com

O. Gudmundsson
e-mail: og@danfoss.com

M. Brand
e-mail: marek.brand@danfoss.com

M. Tunzi
Department of Civil and Mechanical Engineering, University Technical University of Denmark Brovej, Building 118, 2800 Kongens Lyngby, Denmark
e-mail: mictun@dtu.dk

© The Author(s) 2026
D. Vanhoudt (ed.), *Proceedings of the 19th International Symposium on District Heating and Cooling*, Lecture Notes in Networks and Systems 1700,
https://doi.org/10.1007/978-3-032-09844-3_18

1 Introduction

1.1 Background

DH systems are increasingly considered a key asset of sustainable and low emission urban energy strategies, driven by their ability to integrate renewable energy sources and deliver cost-effective thermal energy [1]. Modern DH technology, known as 4GDH, promotes low temperature operation typically with supply and return temperatures around 55 °C and 25 °C respectively, [2]. The 4GDH operating temperatures enhance energy efficiency and facilitate the integration of low-temperature renewable sources, all while maintaining the thermal comfort required for space heating (SH) and domestic hot water (DHW) services. With SH and DHW demands accounting for nearly 80% of European residential energy use the heating sector is a critical target for climate policy and energy system transformation [3]. A common misconception in the transition to 4GDH is that existing buildings must undergo extensive retrofitting to accommodate lower supply temperatures. For example, in Denmark, it has been documented that lower temperatures are achievable [4]. More broadly, lower temperatures are feasible in SH systems during most of the heating season, when not operating under design conditions. During the short periods of higher temperature demands the DH supply temperature can be increased.

DHW systems present a more complex challenge, as the DHW temperature requirements and energy demands stays, while the SH demands are projected to decrease, due to energy renovations and strict building standards for new buildings. Projections suggest the share of DHW could increase from current 15–20% to 40–50% in the near future [5, 6]. This trend, and minimum DHW temperature requirements due to health concerns, positions DHW as a key constraint in efforts to lower DH operating temperatures.

To address the limitations of traditional designs, researchers have investigated, and proven, that integration of electric booster units into DHW substations can effectively be applied for maintaining the temperature of the DHW circulation. Yang et al. [7] examined several configurations of direct electric boosters and Thorsen et al. [8] tested a booster heat pump. However, the use of electricity may increase the overall costs, which generally limits the applicability of these solutions [9].

Therefore, it is essential to investigate new substation designs that can deliver hygienic and comfortable DHW at lower DH supply temperatures while minimizing return temperatures to the DH network without the aid of electricity boosting units.

1.2 Aim

This research explores and compares substation configurations for DHW systems in multi-apartment buildings with DHW circulation loops. The objective is to assess whether these new designs can support low-temperature DH operation. The study evaluates both qualitative and quantitative aspects of the system's performance, focusing on its ability to reduce DH return temperatures at the building level.

2 Method

2.1 Applied Methods

The analysis is based on a quasistatic model of the DH substation concepts, where a relevant range of boundary conditions are explored by a sequence of static simulations performed over the duration of a year. The primary output is the flow weighted DH return temperature under the various boundary conditions or scenarios. The detailed description of the applied method and model is presented in Thorsen et al. [10].

2.2 General Boundary Conditions

The analysis is conducted using weekly intervals over a full year. Annual heat demands are allocated based on the heating degree day (HDD) method, using a base temperature of 17°C. Temperature profiles for the DH network and the building side heating systems are defined in accordance with the characteristics of 4GDH and 3rd generation DH (3GDH) systems, respectively.

The required DHW supply temperature is set to 55 °C. Seasonal variations in DHW demand and circulation losses are assume ±20% and ±10%, respectively. The cold-water inlet temperature is fixed at 10 °C. DHW tapping events are grouped into six-hour daily intervals to represent typical usage patterns. Heat exchangers are dimensioned according to the energy demand of each respective service, SH, DHW and DHW circulation, to ensure accurate thermal performance.

The building heating systems considered include both radiator (RAD) based and underfloor heating (UFH) installations.

2.3 Parallel Concept (Baseline)

The parallel concept as shown in Fig. 1, represent the baseline concept. The upper HEX supplies the DHW service, including both tapping and circulation, while a second HEX, connected in parallel, serves the SH system. This configuration is simple, widely adopted in practice, and characterized by the lowest requirements for components and control complexity among the concepts analyzed.

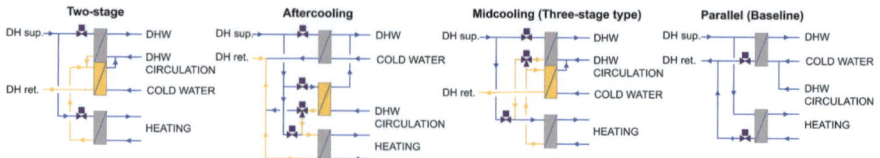

Fig. 1 Analyzed cascaded concepts as well as the baseline concept [11]

2.4 Cascaded Concepts in Focus

Two-Stage Concept

In the two-stage configuration, the HEX for the DHW service is divided into two sections. This arrangement enables a reduction in the primary return temperature from the SH system by using the first section to preheat the cold-water supply for DHW, as illustrated in Fig. 1. The concept requires an additional HEX section, or flow path, for the DHW HEX, compared to the parallel concept.

Aftercooling Concept

The aftercooling configuration adds a dedicated HEX for heating the DHW circulation flow, installed in parallel with the main DHW HEX, as shown in Fig. 1. In this setup, the primary return temperature is further reduced, or aftercooled, by transferring residual heat to the SH system via the SH HEX. The aftercooling concept requires two additional control valves compared to the parallel and two-stage concepts.

Midcooling Concept (Three-Stage Type)

The midcooling concept combines the features of both the two-stage and aftercooling concepts. It enables aftercooling of the primary DHW circulation via the SH, and preheating of the cold-water supply using the primary return flow from the SH system, as illustrated in Fig. 1. The concept requires one additional control valve compared to the parallel and two-stage concept.

2.5 Boundary Conditions, Buildings

The annual DH return temperature depends on the yearly energy consumption of the DHW and SH services, the DH supply temperature, and the characteristics of the employed HEXs. To facilitate a comparative analysis of return temperatures across different applications and building characteristics, two dimensionless parameters were defined: R_1, the ratio of DHW demand to SH demand, and R_2, the ratio of DHW circulation demand to the combined DHW tapping and circulation demands.

The study focused on seven existing Danish multi-apartment buildings, representing typical building stock of varying ages and energy renovation levels. These

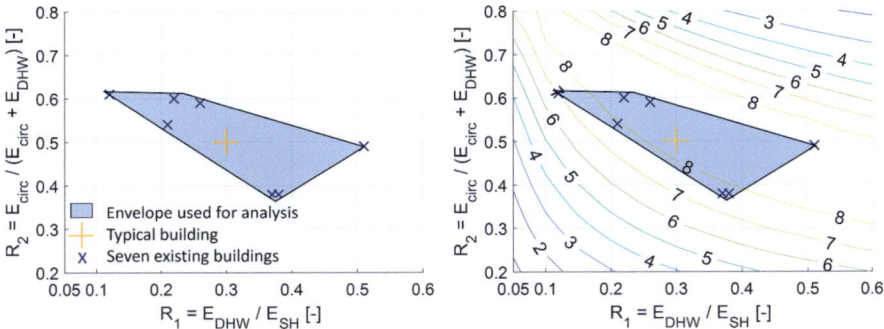

Fig. 2 Contour maps used in analysis. Left: R_1 and R_2 envelope used in the analysis for Copenhagen, covering the seven existing buildings. Blue crosses indicate the existing buildings, and a yellow cross marks a typical building. Right: Contour plot and the envelope shows reduced DH return temperature for the aftercooling concept compared to the baseline concept for the 4GDH and under floor heating case [11]

cases define the range of R_1 and R_2 values considered in the analysis, providing a realistic envelope for evaluating the different DH substation configurations (Fig. 2).

Compared to the Copenhagen climate profile, colder northern climates result in higher SH demands and thus lower R_1 values, while warmer southern climates correspond to higher R_2 values. To account for this climate dependent shift of the R_1/R_2 envelope, adjustments were applied: R_1 envelope values shifted −0.025 for Helsinki, +0.025 for Paris and +0.050 for Rome. The "typical building" shifts accordingly.

2.6 Boundary Conditions, District Heating and Building Space Heating Operating Temperature Profiles

The energy demands of a building are determined by its envelope characteristics and the applied indoor temperature. The resulting DH return temperatures depend on the applied DH supply temperature as well as the configuration of the building's SH and DHW installations. This analysis considers both 3GDH and 4GDH operating temperatures and buildings with either radiators or underfloor heating installations.

For each of these scenarios, the DH supply temperature and the corresponding operating temperatures of the SH systems are defined as functions of the outdoor temperature. These temperature profiles are presented in Fig. 3.

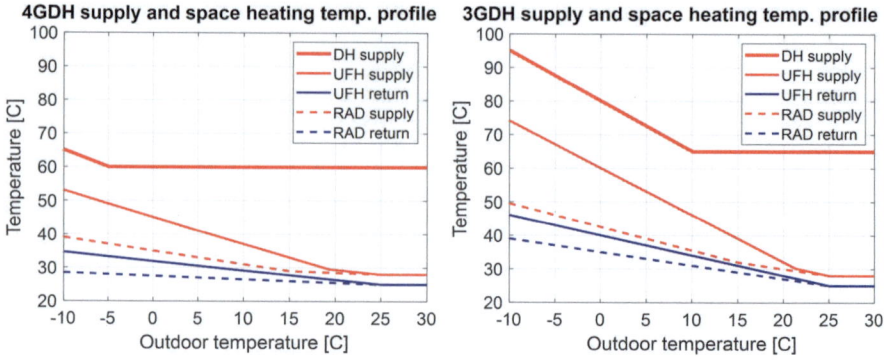

Fig. 3 Temperature profiles, representing 4GDH and 3GDH for UFH and RAD [11]

2.7 Boundary Conditions, Climate Profiles

Four climate zones are considered in the analysis, represented by the design reference year weather profiles for Helsinki, Copenhagen, Paris and Rome. Two key climatic factors influence the performance of the proposed concepts: the duration of the heating season, which determines the potential length of the aftercooling period via the SH circuit, and the outdoor temperature–dependent return temperature of the SH system, which affects the extent of temperature reduction achievable during aftercooling.

3 Results

The results of the analysis are presented pr. substation concept, climate profile and temperature profile, see Fig. 4. The blue bars indicate the range of DH return temperature reductions for the R_1/R_2 building envelope, while the overlaid yellow bars represent the range for the "typical building" case.

3.1 Results for the 4GDH Temperature Profiles

The two-stage concept demonstrates a potential reduction in DH return temperature of approximately 1 to 2 °C, basically independent of climate profile or location. The aftercooling and midcooling concepts exhibit significantly greater performance, with reductions ranging from 3 to 9.5 °C. Among all cases studied, the midcooling concept consistently performs best, explained by its integration of both the two-stage and aftercooling approaches.

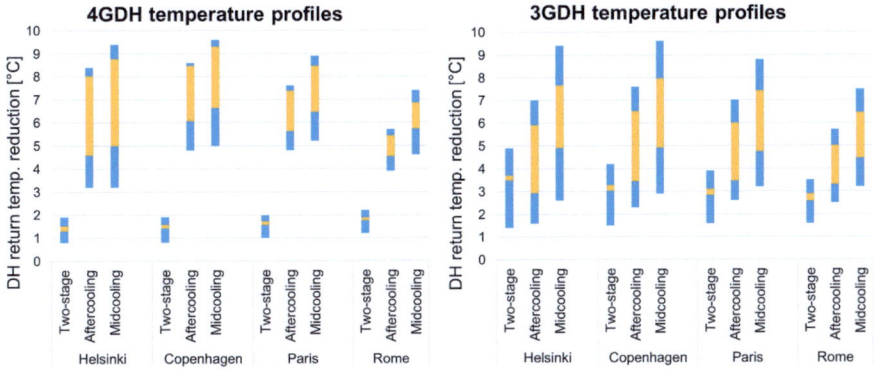

Fig. 4 Analytical results 4GDH and 3GDH [11]

The performance of midcooling and aftercooling has a dependency on the climate profile and location. The relatively long heating season combined with Copenhagen's mild climate results in the highest return temperature reduction potential, between 5 and 9.5 °C. Conversely, Rome's climate leads to the lowest observed performance, between 4 and 7.5 °C.

For the "typical building", the aftercooling and midcooling concepts achieve temperature reductions in the range of 4.5 to 9 °C. Lower values correspond to radiator installations, while higher values are associated with underfloor heating systems.

3.2 Results for the 3GDH Temperature Profiles

The performance of the two-stage concept ranges from 1.5 to 5 °C, with the highest effectiveness observed in colder climates characterized by long heating seasons and in buildings with a high DHW demand, indicated by a high R_1 value. While the lower bound of performance remains relatively constant at approximately 1.5 °C across all climate profiles. The upper bound decreases to around 3.5 °C in the Rome climate, which has the shortest heating season.

Both the aftercooling and midcooling concepts also demonstrate high performance for the 3GDH temperature profiles; however, some reductions are observed compared to the 4GDH profiles. These reductions are primarily attributed to higher building side SH return temperatures, which limit the aftercooling potential through the SH system. As with the 4GDH temperature profile, the performance of both concepts are climate sensitive.

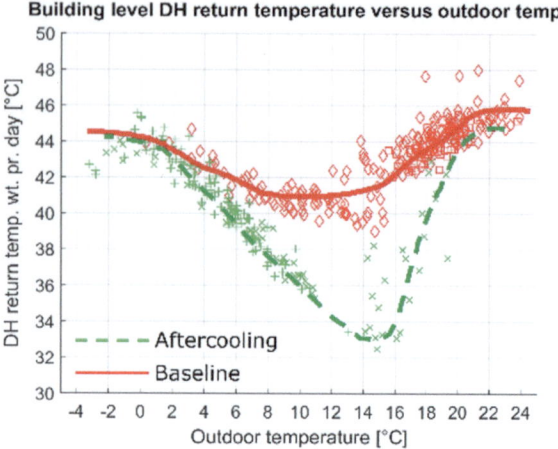

Fig. 5 Field test results, aftercooling concept field test results [11]

4 Field Test Experience

A field test of the aftercooling concept was conducted in Denmark during a two-year period in a multi-apartment building constructed in 1970. Figure 5 presents a comparison between the baseline performance and the aftercooling concept as a function of the outdoor temperature.

During the coldest periods, the reduction in DH return temperature is limited due to the dominance of SH demand; similarly, no reduction occurs in summer when SH is not applied. The greatest temperature reduction is observed at outdoor temperatures between 14 and 16 °C, corresponding to conditions where the DHW circulation primary flow is integrated into the SH circuit. The annual flow-weighted DH return temperature reduction achieved is 3.0 °C, which aligns well with analytical results for the aftercooling concept applied to a building with radiator heating, a 3GDH temperature profile, and Copenhagen's climate, as indicated in Fig. 4, and this verifies our applied method.

5 Discussion and Conclusions

The analysis revealed significant annual flow-weighted DH return temperature reduction potentials, reaching up to 9.5 °C compared to the parallel concept. The midcooling concept demonstrates the highest potential, ranging from 2.5 to 9.5 °C, closely followed by the aftercooling concept, with reductions between 1.5 and 8.5 °C. Both concepts are robust across various climate profiles, represented by Helsinki, Copenhagen, Paris, and Rome, with the best performance observed under long heating seasons and moderate climates. The building system's heating return temperature has a significant influence on performance, making it a key focus area.

For both retrofit and new construction, the midcooling concept is especially suitable when DHW is prepared via an instantaneous heat exchanger. Conversely, when DHW is prepared using a storage tank, the aftercooling concept is more relevant for retrofits, as DHW circulation can be managed through a separate heat exchanger. The aftercooling and midcooling concepts are future-proof solutions, particularly effective under 4GDH temperatures, reducing the DH return temperature by 3 to 8,5 °C and 3 to 9.5 °C, respectively. These analytical results are supported by the two-year field test.

Acknowledgements The preparation of this paper as well as the performed analysis is supported by the European Union's Horizon 2020 research and innovation program under grant agreement no. 101036723, ARV—Climate Positive Circular Communities, the IEA DHC project "Flexibility and DH value chain (FlexVal)" reference XIV-5, as well as EUDP 640232-512102.

References

1. Hansen CH, Gudmundsson O (2018) The competitiveness of district heating compared to individual heating. When is district heating the cheapest source of heating?. Dansk Fjernvarme
2. Lund H et al (2014) 4th Generation District Heating (4GDH). Integrating smart thermal grids into future sustainable energy systems. Energy 68:1–11. https://doi.org/10.1016/j.energy.2014.02.089
3. Eurostat, Energy consumption in EU households. Accessed Jun. 20, 2023. [Online]. Available: https://ec.europa.eu/eurostat/web/products-eurostat-news/-/ddn-20220617-1
4. Østergaard DS, Svendsen S (2018) Experience from a practical test of low-temperature district heating for space heating in five Danish single-family houses from the 1930s. Energy 159:569–578. https://doi.org/10.1016/j.energy.2018.06.142
5. Averfalk H et al (2021) Low-Temperature District Heating implementation guidebook. IEA DHC Report
6. Benakopoulos T, Vergo W, Tunzi M, Salenbien R, Svendsen S (2021) Overview of solutions for the low-temperature operation of domestic hot-water systems with a circulation loop. Energies (Basel), 1–25. https://www.mdpi.com/1996-1073/14/11/3350
7. Yang Q, Salenbien R, Motoasca E, Smith K, Tunzi M (2023) Development and test: future-proof substation designs for the low-temperature operation of domestic hot water systems with a circulation loop. Energy Build 298:113490. https://doi.org/10.1016/j.enbuild.2023.113490
8. Thorsen JE, Svendsen S, Smith KM, Ommen T, Skov M (2021) Feasibility of a booster for DHW circulation in apartment buildings. Energy Rep 7:311–318. https://doi.org/10.1016/j.egyr.2021.08.141
9. Tunzi M, Yang Q, Olesen JB, Diget T, Fournel LC (2024) Economic viability and scalability of a novel domestic hot water substation for 4th generation district heating: a case study of temperature optimization in the Viborg district heating network. Energy 313. https://doi.org/10.1016/j.energy.2024.134010
10. Thorsen JE, Gudmundsson O, Tunzi M, Esbensen T (2024) Aftercooling concept: an innovative substation ready for 4th generation district heating networks. Energy 293:130750, ISSN: 0360-5442, https://doi.org/10.1016/j.energy.2024.130750
11. Thorsen JE, Gudmundsson O, Brand M (2025) Reducing district heating return temperatures by cascading concepts. International conference SZE, June 8–10, 2025, conference presentation, Portoroz, Slovenia

Open Access This chapter is licensed under the terms of the Creative Commons Attribution 4.0 International License (http://creativecommons.org/licenses/by/4.0/), which permits use, sharing, adaptation, distribution and reproduction in any medium or format, as long as you give appropriate credit to the original author(s) and the source, provide a link to the Creative Commons license and indicate if changes were made.

The images or other third party material in this chapter are included in the chapter's Creative Commons license, unless indicated otherwise in a credit line to the material. If material is not included in the chapter's Creative Commons license and your intended use is not permitted by statutory regulation or exceeds the permitted use, you will need to obtain permission directly from the copyright holder.

Status of the VITO Deep Geothermal Project in Mol—Donk (Northern Belgium)

Ben Laenen and Matsen Broothaers

Abstract VITO launched the development of the geothermal plant end 2009 on the assumption that deep geothermal energy can make an important contribution to the energy transition. Over the years, the project generated important data on the local geology, the environmental impact of deep geothermal and the viability of geothermal heat delivery and co-generation. However, VITO did not yet succeed in running the plant at its design capacity. The main question about the plant's operability is the seismic activity. Since the start, two earthquakes were recorded that requested for long-term suspension of operations. The geothermal facilities provide heat to a 3rd generation district heating network and act as Technology Infrastructure for testing innovations to advance deep geothermal energy. In addition, the data that has been collected over the year is available for research and for the development of modelling tools. In this paper we discuss the results of three studies in which the VITO geothermal plant and connected heating network acted as a case to evaluate the impact of demand side management and thermal storage on geothermal district heating. Smart demand side management results in higher thermal output of the plant and more stable production conditions. Thermal storage allows adjusting flow rate to the expected thermal demand and to reducing pump energy. It remains VITO's ambition to use its geothermal facilities as a test site for deep geothermal technologies and to convert the knowledge and experiences gained into tangible value propositions for all stakeholders. Central to this ambition is stable heat delivery in a way that is acceptable for the stakeholders.

Keywords Geothermal energy · District heating · Test facility

B. Laenen (✉) · M. Broothaers
VITO NV, Mol, Belgium
e-mail: ben.laenen@vito.be

M. Broothaers
e-mail: matsen.broothaers@vito.be

1 Project Overview

VITO's deep geothermal project is situated on the VITO Sustainability Park in Mol, Belgium (Fig. 1). The geothermal facilities include several components (Fig. 2):

- Three wells (along hole length or MD): A production well (3610 m MD), an injection well (4311 m MD) and a monitoring well (4905 m MD).
- A building housing the surface installations including installation to process the geothermal brine (Geoloop in Fig. 2), connections to the heating grid, an ORC and facilities to test thermal systems (Thermal applications in Fig. 2), and an adiabatic cooling system (10 × 1 MW cooling capacity). The adiabatic coolers serve as emergency coolers in case the injection temperature rises above 80 °C, to avoid damaging the seals of the injection pumps, and provide cooling for the test facilities and the ORC. The ORC was installed to test the feasibility of geothermal co-generation based on insight from [1, 2].
- A seismic monitoring network consisting of 18 seismometers: 4 at surface, 11 at a depth of 30 to 300 m and 3 placed on the 'bed rock' between 600 and 2052 m.

The geothermal plant is connected to the boiler house of a 3rd generation heating grid that is supplying heat to the research campus of VITO—SCK·CEN and a nearby residential area. Building typologies connected to the network include offices,

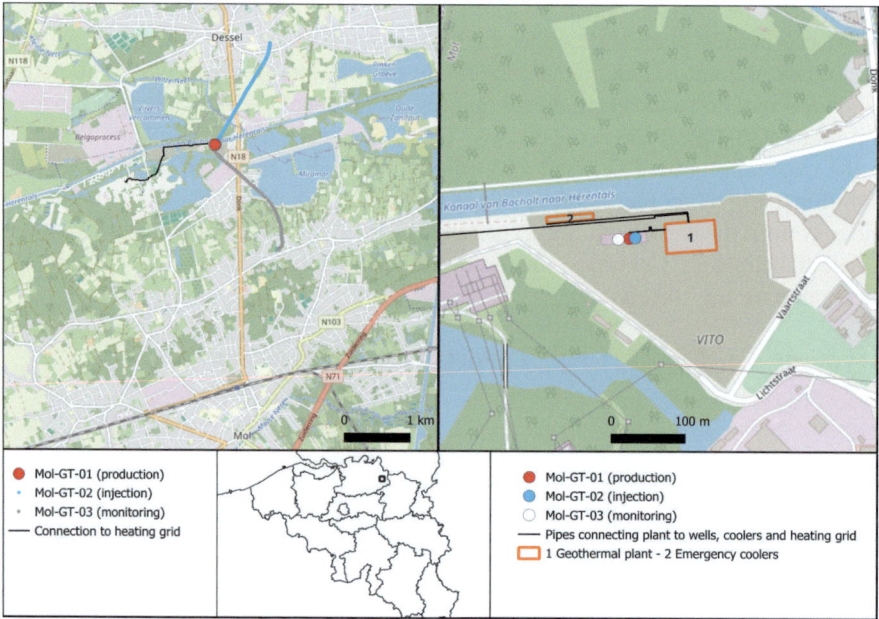

Fig. 1 Overview map with location of the geothermal site and the layout of its components

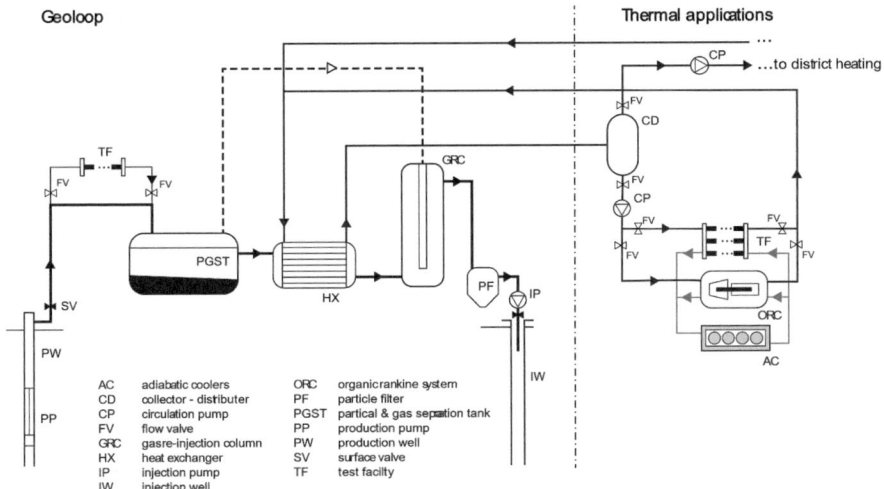

Fig. 2 Schematic of the plant layout from production to injection well

laboratories, industrial halls, apartment buildings, terraced house and detached dwellings.

2 Project History

2.1 Feasibility, Construction and First Start-Up of the Geothermal Plant

VITO started the geothermal project in November 2009. The main goal was to evaluate the techno-economic viability of geothermal district heating and co-generation in Flanders for projects that target carbonate deposits of the Lower Carboniferous Limestone Group. According to previous studies, these carbonates host the largest deep geothermal resources in the northern part of Belgium [3]. An extensive overview of the project is given in Broothaers et al. [4].

After a process of preliminary studies and a 2D seismic exploration survey, the first drilling was spudded on 15 September 2015. Well Mol-GT-01 hit a reservoir at a depth of 3200 to 3400 m in the Lower Carboniferous Limestone Group. Production tests pointed to a productivity of 4 to 5 m^3/h per bar drawdown and a wellhead temperature of 128 °C [5]. Mol-GT-02 was spudded on 2 March 2016. On 23 July 2016, drilling was stopped at a depth of 4341 m (MD). Production tests were not successful. Subsequent injection tests pointed to injectivity of 1.5 to 2 m^3/h per bar injection pressure [5]. Based on these results, it was concluded that circulation at a flow rate up to 150 m^3/h should be possible.

In 2018, Mol-GT-03 was drilled with two objectives: increasing the production capacity to be able to deliver heat to a new heating network in Mol and Dessel and exploring the Devonian sandstones [4]. Mol-GT-03 reached the base of the Lower Carboniferous Limestone Group at 4654 m (MD). Drilling was stopped in the Devonian sandstones at 4905 m MD. As such, the second objective was achieved. Production tests however were unsuccessful, and the well was classified as 'dry'. Consequently, the plans to extend the heating network were shelved. Based on the test results of wells Mol-GT-01 and 02, the geothermal installations were designed for an average flow rate of 100 m^3/h and a production temperature up to 130 °C.

In the spring of 2017, a connecting pipeline between the geothermal site and the boiler house of the heat network was installed. The works started in the first quarter of 2017. During construction of the plant, several technical issues arose that needed solving. This led to a delayed delivering of the building and budget overruns. Major issues were the presence of Naturally Occurring Radioactive Materials (NORM) in the brine [4, 6], proper material selection to reduce the impact of corrosion, defining a proper operational pressure to avoid degassing and minimize scaling and the occurrence of induced seismicity [7].

The plant was completed in November 2018. In the period December 2018–June 2019, a series of production tests were carried out with the aim to start-up stable heat production. The start-up of the plant proved to be difficult. The main reason for this is the high pressure that is required to inject the pumped water back into the reservoir.

Start-up was suspended after an M_L 2.2 earthquake that was felt in Dessel and Mol on 23 June 2019. From July 2019 till December 2020, studies were conducted into the seismic hazard and risks associated with geothermal energy extraction from the Lower Carboniferous limestones [7]. These studies resulted in a research program to determine to what extent deep geothermal energy production in the Mol—Dessel region is technically and economically feasible from a seismological point of view. The research must provide an answer to the question under which conditions the probability of a felt earthquake is acceptably low. It should also improve our knowledge of the relationship between the strength of an earthquake and its effects at surface.

2.2 Production Tests to Define Operative Window

Production was restarted in April 2021 to conduct additional measurements in the context of the seismological research program. From September 2021 till November 2022, the plant has been delivering heat to the boiler house of the heat network [5]. The tests were stopped after a second felt earthquake (M_L 2.1) on 16 November 2022. In the following months, all data gathered since December 2018 were evaluated to decided whether is possible to run the plant with a socially accepted seismologic risk.

In October 2023 VITO invited experts to discuss potential future scenarios for its geothermal facilities. Based on the current understanding and after consultation with multiple partners VITO identified actions to reduce this seismic risk. In November 2023, VITO's board of directors gave permission to adjust the installation and to implement a predictive traffic light protocol with a view to new production tests in winter 2024–2025. Moreover, a 3D seismic campaign was conducted in the fall of 2024. Between November 2024 and June 2025, the plant was running at flow rates between 10 and 20 m^3/h, supplying between 15 and 40% of the demand in the heating grid.

2.3 Production Optimization

Parallel to the seismic research program, Vos evaluated the performance of various parts of the surface installations and looked at options for production optimization [8]. He assessed the effectiveness of choices that were made during construction to mitigate the presence of solids, free gas, corrosion and scaling, and investigated ways to lower the injection pressure and to overcome short power outages or voltage drops.

The assessment led to a series of adjustments to the installations. The issue of free gas in the installations was addressed by increasing the operating pressure to 47.5 bar. Keeping the pressure above the bubble point of the dissolved gases prevents degassing and allows brine and gas to be reinjected together. It also reduces the risk of hammering effects causing fluctuations in injection pressure. Measurements indicated that even with the increased operating pressure, free gas is still formed occasionally, especially during start-up. The removal of this free gas in the surface installations and in the injection well just below the well head, together with the installation of a 100 m long injection tubing, had a positive effect on injection pressure variations.

Analysis of the solids captured by the filters revealed the presence of galena and metallic Pb. Hence, the scaling inhibitor was modified to prevent precipitation of Pb species. The stability of the inhibitor was tested by Pauwels et al. [9].

3 Status

In its current state, the VITO geothermal facilities in Mol qualifies to EARTO's definition of Technology Infrastructure (TI) in the sense that the site *"Consists of facilities, equipment, capabilities, and support services required to develop, test and upscale technology to advance to competitive market entry."*

Besides the hardware, VITO invested in an IT framework to access the production data and the data from the seismometer network in nearly real time (about 1 h delay). The core of the framework is a Python package to easily fetch, update, clean and visualize the production and seismicity data. In addition, geological data about the reservoir (2D and 3D seismic data, test and production data, reservoir model) are available for research and for the development of modelling tools.

4 Research on System Integration and Optimization

4.1 Demand Side Management

In the context of GEOTHERMICA project Heatstore, a Demand Side Management system (DSM) was installed to demonstrate the capabilities of smart control both on the building and network level. For the demonstration the STORM controller [10] was installed in 5 buildings with a combined peak heat load of approximately 2 MW ($\pm$15% of the total peak heat load of the network). The primary target of the DSM is to reduce peak loads and increase the share of geothermal heat delivered to the consumers. The system is online since September 2019. In the following months, response tests were performed to verify whether the control actions had the desired impact on the building's HVAC systems. During these tests the controller was trained.

By actively controlling the demand in the 5 buildings, the return temperature of the network during the heating season could be lowered by 1 to 4 °C (Fig. 3). This results in 2 to 10% higher thermal output of the geothermal plant. Moreover, the return temperature proved to be more stable. This has a positive effect on the injection pressure.

As district heating systems are predominantly demand driven, they are traditionally operated in such a way that heat production is controlled to match the heat

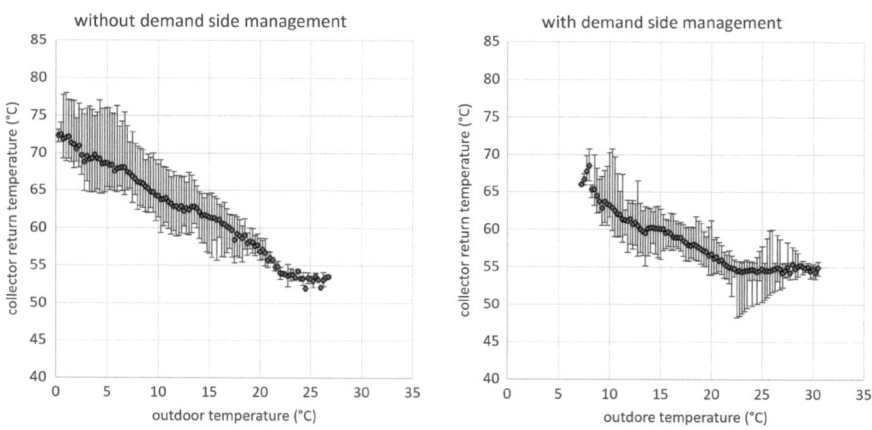

Fig. 3 Return temperature of the heating network as a function of outdoor temperature without and with supply control during the training period of the STORM controller. Bars present the observed temperature range

consumption at any time. Older systems still work on this operating principle. Often, not a lot of emphasis is laid on controlling the demand in such a way that it can benefit production. In 2021, Miglani et al. analyzed the data from the DSM to evaluate a potential optimization utilizing the thermal mass of the connected buildings for storage or additional supply [11]. By simultaneously shifting and managing demand from different buildings, the overall heat demand can be reduced. The results indicate that the total annual flexibility potential in terms of natural gas consumption savings is 610 MWh.

4.2 Thermal Energy Storage

In the context of Horizon 2020 project GeoSmart, the VITO geothermal facilities acted as a virtual case to evaluate the impact of thermal storage on the performance and business case of geothermal district heating [12]. A techno-economic model was set-up based on data from a thermocline storage tank that was installed and tested at the Insheim (DE) geothermal plant. The impact of thermocline storage was evaluated for flow rates in the range of 10 to 75 m^3/h. The low end corresponds to flow rates that were reached during the periods of stable production in 2021–2025. The high end corresponds to low end of the design flow rates. The corresponding thermal output ranges from 300 to 5340 kW, production temperature from 90 to 130 °C.

Over the period 2018–2023, the annual heat demand in the network varied between 20.4 and 25.4 GWh. Base demand is about 800 kW. Peak demand is 12000 kW. The demand on top of the base demand closely correlates with ambient temperature. It is assumed that heat demand will not change significantly in the future. This assumption is justified as the reduction in demand due to measures to increase the energy performance of the buildings that are connected to the network is counterbalanced by new heat demand in the residential area that is connected to the network.

By charging the storage during periods of low demand, the flow rate can be optimized. In case of the geothermal plant in Mol, flow rate has a strong impact on the injection pressure, and hence on the electricity consumption of the injection pumps. To evaluate the impact of the thermocline storage on the performance of the geothermal plant perfect foresight of the heat demand over a period of 12, 24, 36, 48 and 72 h was assumed. The historic demand curves were used to calculate the forecast.

For all flow rates, adding thermocline storage results in a decrease in pump power and an increase in the amount of heat delivered (see Fig. 4). The results show that it is important to tune the size of storage to the output of the geothermal plant and the heat demand in the network. For example, for a flow rate of 50 m^3/h and a storage size of 100 m^3 almost the same amount of heat can be delivered by tuning the flow rate to the expected demand for the next 24 h. As a result, the storage is less charged and discharged. The same is seen at other combinations of flow rate and storage size.

Under the conditions of the VITO geothermal plant, adding a thermocline storage does not result in lower a Levelized Cost of Energy (LCoE) (see Fig. 4). The effect

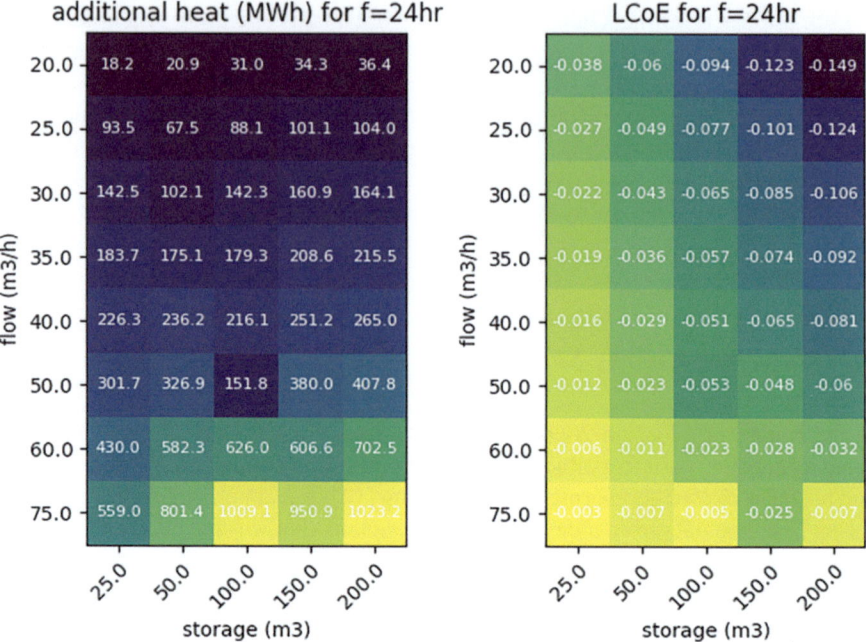

Fig. 4 Impact of thermocline storage on the amount of heat delivered (MWh) and heating cost (€/MWh) of the VITO geothermal plant for a forecasting window of 24 h. The numbers show the difference between a scenario with storage and a scenario without storage

of the additional heat delivery and lower pump energy is counterbalanced by the depreciation of the thermocline storage and additional maintenance costs. For low flow rates, a storage even results in a decline of the net revenues of the plant. This is because the amount of additional heat delivered does not compensate maintenance costs and depreciation. In these cases, it is more effective to tune the flow rate of the geothermal fluid to the forecasted heat demand. At flow rates of 30 m³/h or more, adding thermocline storage can result in higher net revenues and a positive Internal Rate of Investment (IRR). However, careful selection of the storage size is essential to generate additional value. Large storage sizes only pay off for flow rates of 60 m³/h or more.

5 Conclusions

The VITO deep geothermal project in Mol, Belgium, has made significant strides in exploring the potential of deep geothermal energy for heating and co-generation in Flanders and neighbouring regions. Since its inception in 2009, the project has provided valuable data on the geothermal potential of the Lower Carboniferous Limestone Group, the technical and economic feasibility of geothermal energy, and the associated challenges.

The efforts spent in developing the geothermal plant highlight VITO's commitment to advancing geothermal technology and its application in sustainable energy systems. VITO is further developing the facilities. Ongoing adjustments and research are focused on reducing seismic risks and improving the overall feasibility and sustainability of the project with the ambition to create a stable operational environment that is open to third parties for testing innovative technologies for deep geothermal and its integration into next-generation district heating networks.

References

1. Walraven D, Laenen B, D'haeseleer W (2015) Minimizing the levelized cost of electricity production from low-temperature geothermal heat sources with ORCs: water or air cooled? Appl Energy 142. https://doi.org/10.1016/j.apenergy.2014.12.078
2. Van Erdeweghe S, Van Bael J, Laenen B, D'haeseleer W (2019) Optimal configuration for a low-temperature geothermal CHP plant based on thermoeconomic optimization. Energy, 179. https://doi.org/10.1016/j.energy.2019.04.205
3. Berckmans A, Vandenberghe N (1998) Use and potential of geothermal energy in Belgium. Geothermics 27(2):235–242. https://doi.org/10.1016/S0375-6505(97)10010-4
4. Broothaers M, Lagrou D, Laenen B, Harcouët-Menou V, Vos D (2021) Deep geothermal energy in the lower carboniferous carbonates of the Campine Basin, Northern Belgium: an overview from the 1950's to 2020. Zeitschrift Der Deutschen Gesellschaft Fur Geowissenschaften 172(3):211–225. https://doi.org/10.1127/zdgg/2021/0285
5. Bos S, Laenen B (2017) Development of the first deep geothermal doublet in the Campine Basin of Belgium. European Geologist 43:16–20
6. Vasile M, Bruggeman M, Van Meensel S, Bos S, Laenen B (2017) Characterization of the natural radioactivity of the first deep geothermal doublet in Flanders, Belgium. Appl Radiation Isotopes 126:300–303. https://doi.org/10.1016/j.apradiso.2016.12.030
7. Kinscher JL, Broothaers M, Schmittbuhl J, de Santis F, Laenen B, Klein E (2023) First insights to the seismic response of the fractured Carboniferous limestone reservoir at the Balmatt geothermal doublet (Belgium). Geothermics 107:102585. https://doi.org/10.1016/j.geothermics.2022.102585
8. Vos D (2020) Evaluation of production strategies and technologies for long-term production optimization for a Carboniferous carbonate reservoir. INTERREG NWE DGE-ROLLOUT, D.T3.1.2_Evaluation_of_adapted_production_strategies
9. Pauwels J, Salah S, Vasile M, Laenen B, Cappuyns V (2022) Testing the stability of chemical inhibitors at geothermal conditions and their efficiency to prevent galena formation. Geothermics 102:102380. https://doi.org/10.1016/j.geothermics.2022.102380
10. Vanhout D, Claessens B, Desmedt J, Johansson C (2017) Status of the horizon 2020 storm project. Energy Procedia 116:170–179. https://doi.org/10.1016/j.egypro.2017.05.065
11. Miglani S, Allaerts K, Broothaers M (2021) Efficient heat demand management: evaluation of the impact of innovative demand side management on the Balmatt site. INTERREG NWE Deliverable D.T3.2.1
12. GeoSmart: https://www.geosmartproject.eu/repository, last visited on 06/06/2025

Open Access This chapter is licensed under the terms of the Creative Commons Attribution 4.0 International License (http://creativecommons.org/licenses/by/4.0/), which permits use, sharing, adaptation, distribution and reproduction in any medium or format, as long as you give appropriate credit to the original author(s) and the source, provide a link to the Creative Commons license and indicate if changes were made.

The images or other third party material in this chapter are included in the chapter's Creative Commons license, unless indicated otherwise in a credit line to the material. If material is not included in the chapter's Creative Commons license and your intended use is not permitted by statutory regulation or exceeds the permitted use, you will need to obtain permission directly from the copyright holder.

Proposal for a Method to Simultaneously Maximize the Economic and Environmental Values of a District Heating and Cooling System for an Electricity Market

Kohei Tomita, Yutaka Iino, and Yasuhiro Hayashi

Abstract District heating and cooling system (DHC), which is responsible for supplying heat, cold, and electricity to the region, often have the ability to generate electricity as well as generate heat and cold. In our previous study, we proposed an operational method that would enable DHC to simultaneously supply electricity to the electricity market in Japan while fulfilling its responsibility for local demand, as well as a method for evaluating its economic efficiency. This study focuses on the issue that environmental value is not yet fully evaluated in many electricity markets around the world and proposes a method for determining an operation that maximizes the total of the economic value and environmental value of DHC. In addition, authors examine the possibility of changes in the optimal DHC equipment configuration, operation methods, and conditions through sensitivity analysis of carbon prices and CO_2 emission factors. Based on these research results, conditions under which it will be necessary to update the DHC equipment configuration and operation methods are organized in future fluctuation scenarios of carbon price and CO_2 emission factors.

Keywords District heating and cooling system · Combined heat and power · Electricity market · Mixed integer nonlinear programming method

1 Introduction

District heating and cooling system (DHC), which is responsible for supplying heat, cold, and electricity to the region, often have a combined heat and power system (CHP) that has the ability to generate electricity as well as generate heat and cold. Hence, they have high potential as a distributed energy source that can contribute to

---

K. Tomita (✉) · Y. Iino · Y. Hayashi
Graduate School of Advanced Science and Engineering, Waseda University, Shinjuku, Japan
e-mail: kohei.tomita@fuji.waseda.jp

the stabilization of the power system by utilizing the CHP. In recent years, the establishment of electricity markets that procure power sources from various providers is being completed or considered in countries around the world. In our previous study, we proposed an operational method that would enable DHC to simultaneously supply electricity to the electricity market in Japan while fulfilling its responsibility for regional heat, cold and electricity demand, as well as a method for evaluating its economic efficiency [1–4].

On the other hand, even though one of the reasons for establishing electricity markets was to ensure enough electricity to absorb the fluctuation of large amounts of variable renewable energy for decarbonization, many electricity market bidding conditions around the world depend on economic value and do not sufficiently consider environmental value. By considering both economic value and environmental value simultaneously, it is possible that the power sources auctioned in the electricity market and their operating methods may change. On the other hand, to the best of our knowledge, there are no previous studies that simultaneously evaluate the economic value and environmental value of DHC in the electricity market and examine operation methods.

In this study, we focus on DHC and propose a method for determining an operation that maximizes the total of the economic value composed of operation cost and environmental values composed of CO_2 emission from fuel and grid power consumption. Finally, based on the method, we conduct a sensitivity analysis of changes in DHC operation using multiple scenarios of carbon prices and CO_2 emission factors, and discuss the environmental value of DHC and the future direction of the electricity market. The novelty of this study lies in proposing a method for determining operations that simultaneously optimize economic value and environmental value in the electricity market, and in deriving the conditions under which it is necessary to update the optimal equipment configuration and operation methods of DHC in future fluctuation scenarios of carbon price and CO_2 emission factors.

2 Operation of a DHC for the Electricity Market

The normal operation and the operation for the electricity market are defined as follows. Figure 1 shows the changes in DHC power generation methods of each operation.

2.1 Normal Operation

DHC operates to meet the heat, cold and electricity demand of a particular region. Therefore, normal operation is not affected by the electricity market (see Fig. 1 left).

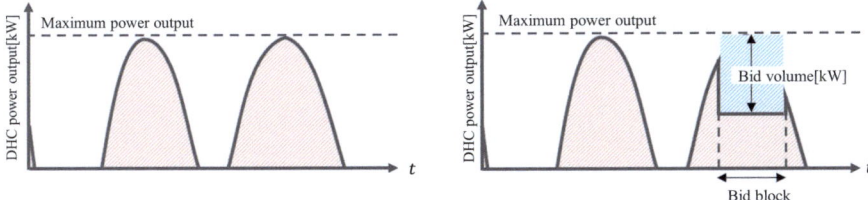

Fig. 1 Power output image of DHC (Left figure: Normal operation, Right figure: Market standby operation)

2.2 Market Standby Operation

DHC operates to ensure sufficient excess power output capacity equal to or greater than the bid volume kW in the bid block to respond to power supply commands from the electricity market while meeting the heat, cold and electricity demand of a particular region (see Fig. 1 right).

3 Modeling a DHC

3.1 DHC Model

This study cites the same DHC model as our previous research [1–4]. This DHC existed in Japan (Specific locations and names cannot be disclosed under NDA.). The maximum power output of the DHC is 15,700 kW, which is provided by gas engine generator No. 4, which is part of the CHP (equipment Nos. 4, 5, and 6). Electricity from the CHP and purchased electricity from the external power system will be used to supply electric turbo chiller No. 8 and meet regional electricity demand. Furnace flue boiler No. 3 and waste heat boiler No. 5 meet regional heat demand. Steam turbine turbo chiller No. 2, hot water absorption chiller No.6, steam absorption chiller No.7 and electric turbo chiller No. 8 meet regional cold demand (see Fig. 2). Details of DHC equipment information are shown in Table 1. Regional demand is based on summer weekday data, with heat and cold demand shown in Fig. 3 and electricity demand in Fig. 4. Electricity demand includes regional electricity demand and an internal load of approximately 4,000 kW per hour.

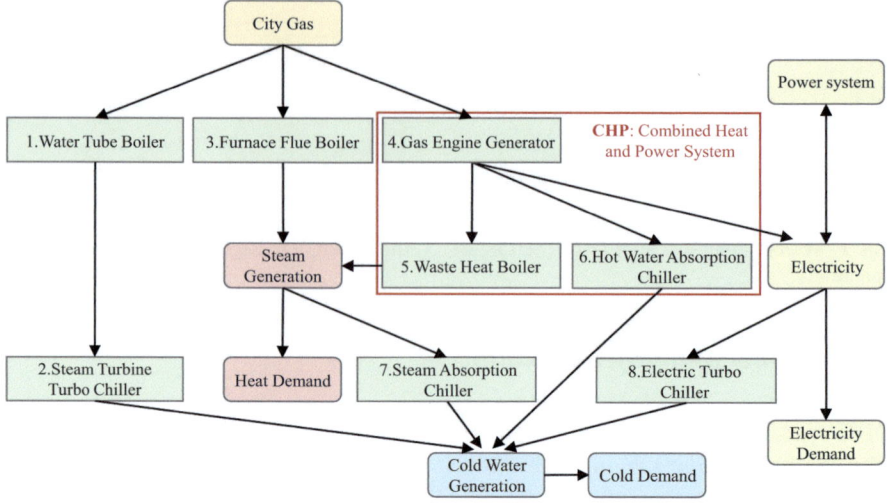

Fig. 2 DHC model

Table 1 DHC equipment information

Equipment number and name	Lower output limit	Upper output limit
1. Water tube boiler	65 GJ/h	260 GJ/h
2. Steam turbine turbo chiller	63 GJ/h	253 GJ/h
3. Furnace flue boiler	14.4 GJ/h	125.2 GJ/h
4. Gas engine generator	3,485 kW	15,700 kW
5. Waste heat boiler	6.0 GJ/h	26.8 GJ/h
6. Hot water absorption chiller	2.2 GJ/h	9.0 GJ/h
7. Steam absorption chiller	9.5 GJ/h	75.6 GJ/h
8. Electric turbo chiller	4.4 GJ/h	21.4 GJ/h

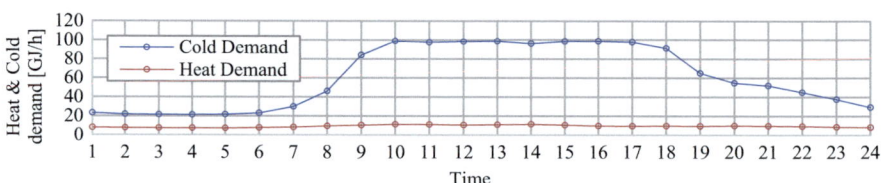

Fig. 3 Heat and cold demand

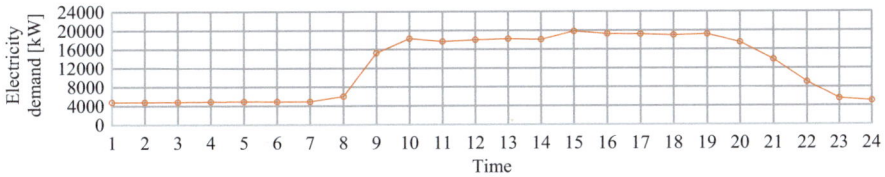

Fig. 4 Electricity demand

3.2 Formulation of DHC Model

As a method for determining an operation that maximizes the total of the economic value composed of operation cost and environmental values composed of CO_2 emission from fuel and grid power consumption, the DHC model is formulated as follows. Details of each parameter are shown in Table 2.

Objective function

The objective function for minimizing costs considers not only operating costs but also carbon prices to maximize the total of the economic value and environmental values.

Table 2 Parameters information

Parameters	Description
$Cost$	Operating costs [yen]
C_g, C_p	Gas price [yen/m^3], Electricity price [yen/kWh]
C_{co2}	Carbon price [yen/kg-CO_2]
$\varepsilon_g, \varepsilon_p$	Gas carbon emission factor [kg-CO_2/m^3], Electricity carbon emission factor [kg-CO_2/kWh]
$g(t), p(t)$	Purchased gas volume [m^3], Purchased electricity volume [kWh]
p_{MAXout}	Maximum power output capacity (Equipment No.4) [kW]
cap_{target}	Bid power [kW]
ts	Start time of the bid time block
$s_g(t), cw_g(t)$	Steam generation volume [GJ/h], Cold water generation volume [GJ/h]
$s_d(t), cw_d(t), p_d(t)$	Steam demand [GJ/h], Cold water demand [GJ/h], Electricity demand [kWh]
i	Equipment number
$x_i(t), y_i(t)$	Input of each equipment, Output of each equipment
s_i	ON–OFF status of each equipment (0 or 1)
$\delta_{i,up}(t), \delta_{i,down}(t)$	Start-Stop switch of each equipment (0 or 1)
u	Maximum number of starts/stops (=4 times)
T_{up}	Minimum continuous operating time [h] (= 8 h)

$$Cost = \sum_{t}^{T}\{(C_g + C_{co2} \times \varepsilon_g) \times g(t) + (C_p + C_{co2} \times \varepsilon_p) \times p(t)\} \rightarrow Minmize \tag{1}$$

Input/output functions of each equipment

The input and output functions of each equipment are nonlinear, based on the characteristics of the actual devices.

$$y_i(t) = f_i(x_i(t)) \tag{2}$$

Regional heat, cool, and electricity demand constraints

$$s_g(t) = y_3(t) + y_5(t) - x_7(t) \geq s_d(t) \tag{3}$$

$$cw_g(t) = y_2(t) + \sum_{i=6}^{8} y_i(t) \geq cw_d(t) \tag{4}$$

$$p(t) + y_4(t) = p_d(t) \tag{5}$$

Market standby constraint

$$p_{MAXout} - \sum_{t=ts}^{ts+2} y_4(t) \geq cap_{target} \tag{6}$$

Num of start/stop cycles and minimum continuous operation time constraints

$$\delta_{i,up}(t) + \delta_{i,down}(t) \leq 1 (i=1,2,\ldots 8) \tag{7}$$

$$s_i(t) - s_i(t-1) = \delta_{i,up}(t) - \delta_{i,down}(t)(i=1,2,\ldots 8) \tag{8}$$

$$\sum_{t=1}^{T} \delta_{i,up}(t) + \delta_{i,down}(t) \leq u(i=1,2,\ldots 8) \tag{9}$$

$$\sum_{j=t}^{t+T_{up}-1} s_i(j) \geq T_{up}\delta_{i,up}(j)(i=1,2,\ldots 8) \tag{10}$$

The sensitivity analysis in the following chapters is simulated based on this formulation using mixed-integer nonlinear programming.

Table 3 Conditions on electricity market

Condition items	Description
Target market	Electricity market in Japan (Balancing market)
Bid power	8,000 kW
Bid time block	12:00–15:00

4 Conditions for Numerical Simulation

In Japan, electricity is generally procured through bilateral contracts with utility companies or electricity markets, and gas is procured through bilateral contracts. As a premise of this study, DHC generates heat and cold for the region using electricity and gas procured at fixed prices under bilateral contracts, while simultaneously selling electricity on the electricity market using surplus power generation capacity of CHP.

4.1 Conditions on Electricity Market

This study targets the Japanese electricity market which is called balancing market. In this market, bidding takes place in eight blocks of three hours each day, with a capacity of 1,000 kW or more. The bid conditions for the simulation are as shown in Table 3.

4.2 Conditions on Prices and Carbon Emission Factor

The conditions on prices and carbon emission factor for this simulation are as shown in Table 4. Gas and electricity prices are based on those of typical utility companies in Japan. Carbon price refer to the 2030 carbon price in the IEA report's Net Zero Scenario for developed countries [5]. (The exchange rate is 145 yen/USD.) Carbon emission factors refer to actual values for gas and electricity in fiscal 2023 reported by Tokyo Gas Co., Ltd. [6] and Tokyo Electric Power Company Energy Partners, Inc. [7]. In addition, four cases of carbon emission factor combinations are prepared based on whether or not CO_2 emissions from gas and electricity were considered, either individually or both.

Table 4 Conditions on prices and carbon emission factor

Condition items	Description
Gas price C_g	90 [yen/m^3]
Electricity price C_p	17 [yen/kWh]
Carbon price C_{co2}	20.3 [yen/kg-CO$_2$]
Carbon emission factor Gas ε_g, electricity ε_p	Case 0: ε_g = None, ε_p = None Case 1: ε_g = None, ε_p = 0.408 [kg-CO$_2$/kWh] Case 2: ε_g = 2.05 [kg-CO$_2$/m^3], ε_p = None Case 3: ε_g = 2.05 [kg-CO$_2$/m^3], ε_p = 0.408 [kg-CO$_2$/kWh]

5 Results and Discussion

The comparison results for each case simulation are shown in Table 5. Only Case 2, which considered only gas carbon price, showed a difference in operation compared to Case 0. Figure 5 shows operation of the major equipment in DHC for each case. In Case 2, operation shifted to electricity-centric mode due to the impact of gas carbon price. DHC achieves high efficiency through an optimal balance of electricity and gas usage, so the ratio of electricity and gas prices is considered an essential parameter that affects operation methods of DHC.

In other words, the balance between the price of gas $C_g + C_{co2} \times \varepsilon_g$ and the price of electricity $C_p + C_{co2} \times \varepsilon_p$ in the objective function is an important factor in determining DHC operation. For further investigation, Fig. 6 shows a sensitivity analysis of purchased gas and electricity volume of DHC in response to changes in each price. DHC operation will remain unchanged (White zone in the figure) as long as a certain balance between electricity and gas prices is maintained. However, if prices become skewed, DHC operation will switch to gas-centric or electricity-centric modes (Blue/Red zones in the figure).

The introduction of renewable energy and the replacement of gas with hydrogen may cause fluctuations in carbon prices and CO$_2$ emission factors in the future. Based on the results of this study, which showed that environmental values can change operation methods and cost, it is necessary to continuously update the optimal operation and configuration of DHC that simultaneously considers environmental and economic values. Furthermore, considering the results of this study and the increasing importance of environmental values in recent years, it is essential to discuss the

Table 5 Comparison of operation and costs for each case

Case	Difference in operation	Difference in operation cost (%)
Case 0 (Comparison standard)	\	\
Case 1	No	+13.5
Case 2	Yes	+28.1
Case 3	No	+46.1

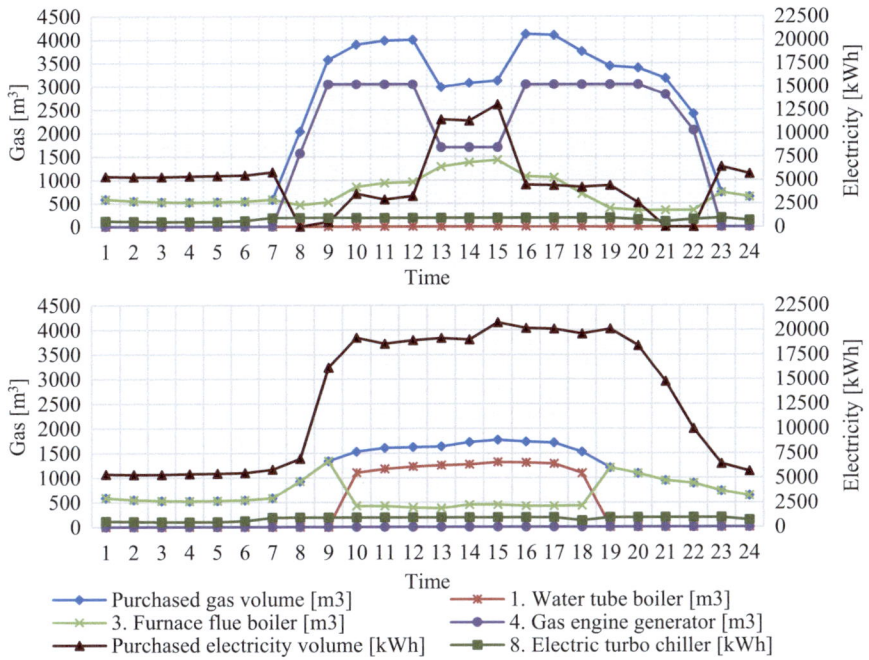

Fig. 5 DHC operation (Upper figure: Case 0/1/3, Lower figure: Case 2)

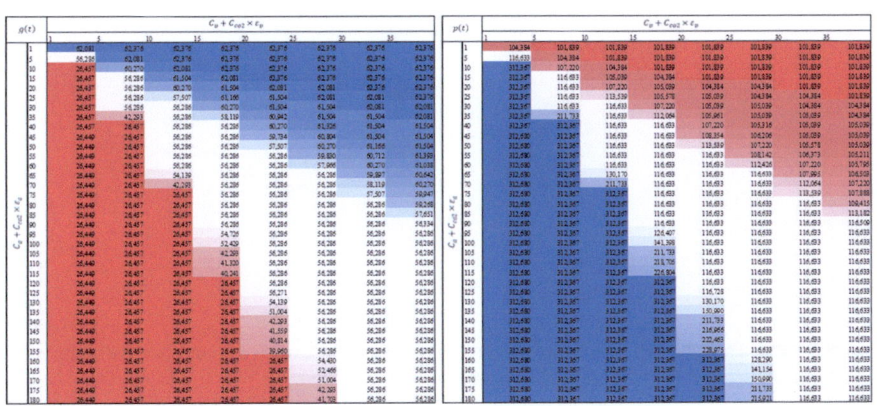

Fig. 6 Sensitivity analysis (Left figure: Purchased gas volume per day [m^3], Right figure: Purchased electricity volume per day [kWh])

appropriate system of electricity markets in the future, which currently evaluate mainly economic values.

References

1. Ito M, Takano A, Shinji T, Yagi T, Hayashi Y (2017) Electricity adjustment for capacity market auction by a district heating and cooling system. Appl Energy 206:623–633
2. Tomita K, Ito M, Hayashi Y, Yagi T, Tsukada T (2018) Electricity adjustment by aggregation control of multiple district heating and cooling systems. Energy Procedia 149:317–326
3. Tomita K, Iino Y, Hayashi Y, Yamamoto Y, Kobayashi K (2020) Partitioning method for the large-scale operation planning problem of a district heating and cooling system for electricity adjustment. IEEJ Trans Power Energy 140:94–103
4. Tomita K, Iino Y, Hayashi Y, Yamamoto Y, Kobayashi K (2020) Evaluation and visualization of kW/kWh cost of a district heating and cooling system for electricity adjustment. In: 2020 4th International conference on green energy and applications (ICGEA)
5. IEA (2023) Net zero roadmap a global pathway to keep the 1.5 °C goal in reach
6. Ministry of the Environment (2023) CO_2 emission factors by gas Company; https://ghg-santei kohyo.env.go.jp/calc (Last accessed on May 31, 2025)
7. Tokyo Electric Power Company Energy Partners, Inc.: CO_2 emission factor for fiscal year 2023 (2024)

Open Access This chapter is licensed under the terms of the Creative Commons Attribution 4.0 International License (http://creativecommons.org/licenses/by/4.0/), which permits use, sharing, adaptation, distribution and reproduction in any medium or format, as long as you give appropriate credit to the original author(s) and the source, provide a link to the Creative Commons license and indicate if changes were made.

The images or other third party material in this chapter are included in the chapter's Creative Commons license, unless indicated otherwise in a credit line to the material. If material is not included in the chapter's Creative Commons license and your intended use is not permitted by statutory regulation or exceeds the permitted use, you will need to obtain permission directly from the copyright holder.

Optimizing the Next Generation of District Heating and Cooling Systems While Ensuring Reliable Domestic Hot Water Supply

Mohammad-Reza Kolahi● and Martin K. Patel●

Abstract The shift toward next-generation district heating and cooling networks (characterized by low- and ultra-low temperature grids) offers improved energy efficiency and greater potential for renewable integration. However, these systems face challenges in reliably supplying domestic hot water (DHW), which requires higher temperatures to ensure user comfort and hygienic safety. This study evaluates four scenarios for delivering both space heating (SH) and DHW in a district system, using optimization to determine the optimal supply and return temperatures that minimize total electricity consumption. Scenario 1 uses electric boilers for DHW; Scenario 2 employs a single decentralized heat pump for both SH and DHW; Scenario 3 introduces a booster heat pump for DHW; and Scenario 4 incorporates thermal storage with scheduled DHW production. Results show that Scenarios 2 and 3 significantly reduce electricity consumption compared to the others, with Scenario 3 achieving the lowest annual demand (520 [MWh/year]). Duration load curves also demonstrate improved load distribution, particularly in Scenario 3. The findings underscore the trade-offs between energy efficiency, system complexity, and hygienic safety, and suggest that further optimization of storage operation and comprehensive cost–benefit analyses will be essential in future planning.

Keywords District heating and cooling · Low-temperature grid · Domestic hot water · Heat pump · Booster · Thermal energy storage

1 Introduction

New generations of district heating networks (DHNs), commonly referred to as 4th generation networks (low-temperature) and 5th generation networks (ultra-low temperature, also known as Thermal Source Networks [1]), have been designed to

---

M.-R. Kolahi (✉) · M. K. Patel
Group for Energy Efficiency, Institute for Environmental Sciences and Department F., A. Forel for Environmental and Aquatic Sciences, University of Geneva, Geneva, Switzerland
e-mail: Mohammadreza.Kolahi@unige.ch

supply heat for both space heating (SH) and domestic hot water (DHW) at lower temperatures and for providing cooling in the case of Thermal Source Networks (TSNs). This approach aims to improve overall system efficiency and facilitate the integration of renewable energy sources.

Operating DHNs at low temperatures presents a key challenge: domestic hot water typically requires higher temperatures than space heating, both for user comfort and to meet hygiene standards, particularly for preventing Legionella growth. However, if the DHN is forced to maintain high supply temperatures continuously in order to satisfy DHW demand, system efficiency is compromised. The coefficient of performance (COP) of the central heat pump decreases, and thermal losses in the distribution network increase significantly.

Various approaches have been introduced in the literature to address this challenge, such as specialized configurations of heat pumps or different substation circuits designed for DHW preparation. Other authors suggest integrating substations with supplementary heat sources, including renewable energy systems or booster heat pumps, to meet DHW temperature requirements. For example, Zhu et al. [2] investigated the integration of booster heat pumps in ultra-low temperature DHNs for supplying both SH and DHW. However, their configuration includes decentralized heat pumps only for DHW, not for SH, as they assumed that space heating would be directly supplied by the high-temperature district heating grid. Other studies (e.g., Bordignon et al. [3] and Barbu et al. [4]) explored the use of solar thermal or photovoltaic systems to support DHW generation. While promising, these approaches may be less effective in regions with limited solar availability due to climate conditions.

The higher temperature requirement for DHW is not only for user comfort, it is primarily a hygienic necessity to mitigate the risk of Legionella bacteria growth. This risk also strongly depends on the system type. In instantaneous (tankless) systems, water is heated on demand via a heat exchanger, heat pump, or electric boiler, without significant storage. Because the volume of warm water is minimal, typically less than 3 L from heater to tap, the risk of Legionella proliferation is negligible [5]. These systems can safely operate at lower DHW supply temperatures. In contrast, storage tank DHW systems maintain a larger volume of warm water, often with continuous circulation. Due to the potential for stagnation and prolonged exposure to suboptimal temperatures, these systems require sustained higher temperatures to ensure Legionella safety [5].

This study aims to compare several scenarios for meeting both space heating and domestic hot water demands without the need to maintain high supply temperatures in the district heating network at all times. Each scenario is evaluated under optimized operating conditions, with the objective of minimizing the total electricity consumption of the DHN system. The scenarios considered are:

- Scenario 1: Decoupled DHW, individual electric boilers in each building
- Scenario 2: Instantaneous heat pump system for simultaneous SH and DHW
- Scenario 3: Separate decentralized booster heat pump for DHW
- Scenario 4: Timed DHW production with On-Site thermal storage.

2 System Description

2.1 DHN Description

The new generation of DHN systems is characterized by low-temperature operation and the integration of advanced heat pump technologies. In the configuration considered, a centralized air-source heat pump functions as the primary heat source. The centralized air-source heat pump extracts ambient heat from the outdoor air to supply energy and control the temperature of the DHN. In addition, multiple decentralized water-to-water heat pumps are installed at the building level to meet individual heating demands. The working fluid circulating in the network's loop is water, drawn from and returned to the DHN circuit. The supply and return temperatures of the DHN are optimized to minimize the total electricity consumption of the system. Based on the hourly profiles of the buildings' SH and DHW demands, the optimization problem was formulated with the objective of minimizing the total electricity consumption of the system. The supply and return temperatures of the network (over the course of one week) were defined as the optimization variables. A Particle Swarm Optimization (PSO) algorithm was employed to solve the problem. The model iteratively adjusted the network's temperatures by integrating dynamic demand profiles, ambient temperature data, and detailed heat pump performance characteristics, ensuring that hourly demands were met while determining the optimal operating parameters.

The TESSA tool [6], is utilized to define the DHN topology and collect essential input data. TessaOSA (Thermal Energy System Simulation Assistant) [6] offers a comprehensive building database that includes parameters such as building type, construction year, energy demand, and heated area. In addition, two advanced models are developed: the first simulates the performance of residential heat pumps, and the second optimizes the overall DHN design based on the detailed thermal and operational modeling. This integrated approach enables an accurate evaluation of system performance under varying demand and ambient conditions.

In this study, a real district is used to calculate key parameters across different scenarios, enabling a consistent comparison of results within a single case study. The selected case is located in Grandvaux, a village in the Swiss canton of Vaud, consisting of 99 buildings situated in a suburban setting. Figure 1 presents the schematic layout of the district heating system and the corresponding annual thermal load curve, both generated using the TESSA tool.

2.2 Scenarios

In this study, four scenarios are evaluated to compare different approaches to domestic hot water production within a low-temperature DHN, while meeting both SH and DHW demands with minimum electricity consumption.

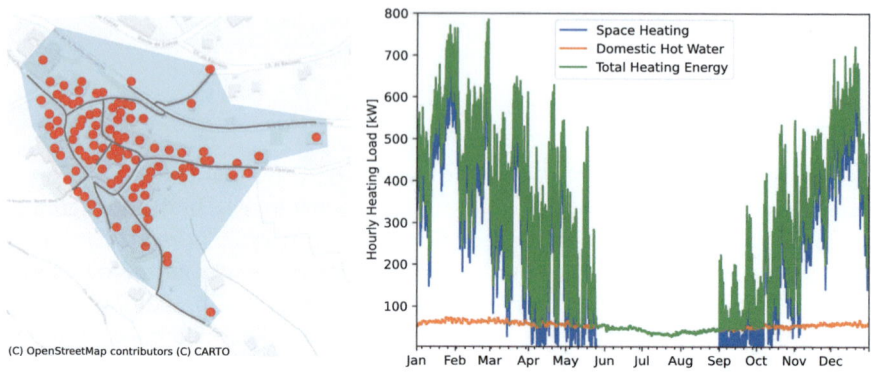

Fig. 1 District heating network system schematic with load curve for Grandvaux, Vaud

In Scenario 1 (Decoupled DHW), individual electric boilers are installed in each building to produce DHW independently of the district heating network. This configuration fully decouples DHW production from the DHN, allowing the network to operate only for SH requirements using one decentralized heat pump in each building.

In Scenario 2 (Instantaneous heat pump system for simultaneous SH and DHW), a single decentralized water-to-water heat pump in each building provides both SH and DHW at the same time. Hot water is generated on demand through an instantaneous tankless system, eliminating the need for DHW storage. This setup improves flexibility and reduces standby losses while maintaining comfort and hygiene.

In Scenario 3 (Separate decentralized booster heat pump for DHW), in addition to the decentralized heat pump for SH, a separate water-to-water booster heat pump is installed in each building specifically for DHW production. This unit boosts heat from the return pipe of the space heating circuit. In cases where no SH is needed, the DHW booster connects directly to the DHN supply pipe. Figure 2 illustrates the schematic configuration of this scenario for both cases, whether space heating is active or not.

Finally, in Scenario 4 (Timed DHW production with on-site thermal storage), DHW is generated at predefined intervals and stored in a thermal storage tank within each building. In this scenario, a single decentralized heat pump is installed for SH, and during specific intervals, the same heat pump also generates DHW for storage. In this work, parameters such as charging/discharging schedules and storage tank sizes are not optimized. It is assumed that the storage tanks are charged for one-third of each day, during off-peak hours.

In all scenarios, the system is optimized to deliver both SH and DHW with the goal of minimizing total electricity consumption, considering the interactions between the centralized and decentralized components.

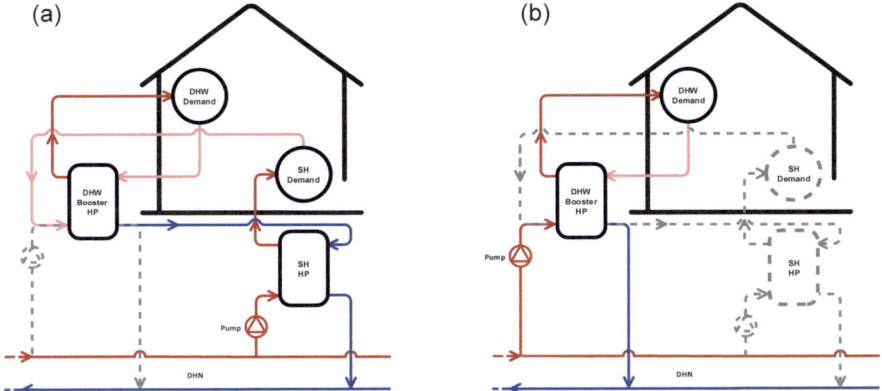

Fig. 2 Schematic configuration of Scenario 3: **a** when both SH and DHW are required, and **b** when only DHW is needed

2.3 Legionella Problem and DHW Temperatures

Many European health regulations allow for lower domestic hot water temperatures in tankless systems. For example, the German standard DVGW W551 states that if the total volume of hot water between the generation point and the tap is less than 3 L, the risk of Legionella proliferation is effectively eliminated [5]. Similarly, Denmark's building code (DS 439) mandates a minimum DHW temperature of ~45 °C at the tap for systems with limited volume [7], while other countries such as Sweden and Switzerland require a minimum of around 50 °C at the point of use [7, 8]. In contrast, DHW systems incorporating storage tanks must maintain higher temperatures to prevent bacterial growth. Most engineering standards and public health guidelines across Europe recommend that stored hot water be maintained at or above 60 °C. For instance, Swiss regulations (SIA 385/1) require a minimum of 60 °C at the outlet of water heaters [8]. Similarly, the Swedish building code specifies at least 50 °C throughout the DHW system [7], while Norway adopts even stricter safety margins [5]. These requirements are grounded in microbiological research showing that Legionella bacteria exhibit virtually no growth at temperatures ≥ 55 °C and are rapidly inactivated at ~60–70 °C [5].

Accordingly, in this study, the DHW temperature is set to 50 °C for the tankless configurations (Scenarios 1, 2, and 3), where the Legionella risk is negligible due to the minimal water volume. For Scenario 4, which includes thermal storage, the DHW temperature is set to 65 °C to ensure Legionella safety under all operating conditions.

3 Results and Discussion

In this section, the results of modeling the district heating network for all scenarios are presented. Figure 3 shows the district's supply and return temperatures throughout the weeks of a typical year under the four scenarios. All temperatures are optimized such that the total electricity consumption of the district is minimized.

As seen in Fig. 3, in Scenario 1, during the summer, when the ambient temperature is high and there is no space heating demand, the district heating network is turned off because the domestic hot water demand is decoupled from the grid. However, there are three weeks when the average outside temperature drops below the heating temperature limit, causing the DHN to become active. A clear relationship can also be observed between the ambient temperature and the grid's supply and return temperatures: the lower the ambient temperature, the higher the supply temperature and the temperature difference (between supply and return).

In the graph for Scenario 2 from Fig. 3, both SH and DHW are instantaneously produced from a single decentralized heat pump (connected to the grid) in each building. As a result, the supply temperature of the grid is generally high, except during periods with no SH demand. Since the required DHW temperature is typically higher (and fixed) compared to SH temperature, the grid's supply temperature does not strongly correlate with ambient temperature. However, the return temperature remains highly dependent on the outside temperature.

In Scenario 3 from Fig. 3, the use of a booster heat pump for DHW increases the grid's supply temperature compared to Scenario 2. As the booster heat pump is connected to the SH heat pump (see Fig. 2), more heat must be extracted from the SH heat pump to provide the required boost for DHW. Therefore, the optimal supply temperature is higher to minimize total electricity consumption.

Finally, in the graph for Scenario 4 from Fig. 3, the use of thermal storage also keeps the grid's supply temperature high for most of the year. This is particularly evident because the required DHW temperature in this scenario is 65 °C, leading to a higher supply temperature compared to Scenario 2. Similar to Scenario 2, the return temperature of the grid remains highly dependent on the outside temperature.

Table 1 presents the total electricity consumption of the district system across the different scenarios. As expected, Scenario 1 results in the highest electricity consumption (910 [MWh/year]), since using an electric boiler instead of a heat pump is an inefficient method for producing DHW. Scenario 4 also shows relatively high consumption (569 [MWh/year]), due to the use of thermal storage, which not only requires a higher supply temperature but also incurs significant thermal losses. Scenario 2 achieves a slightly lower consumption level (559 [MWh/year]), as heat pumps can produce heat with a high coefficient of performance (COP), consuming less electricity compared to electric boilers. However, Scenario 3 yields the lowest overall electricity consumption (520 [MWh/year]). This is due to the use of a booster heat pump for DHW, which enables more efficient heat utilization and further reduces total energy use.

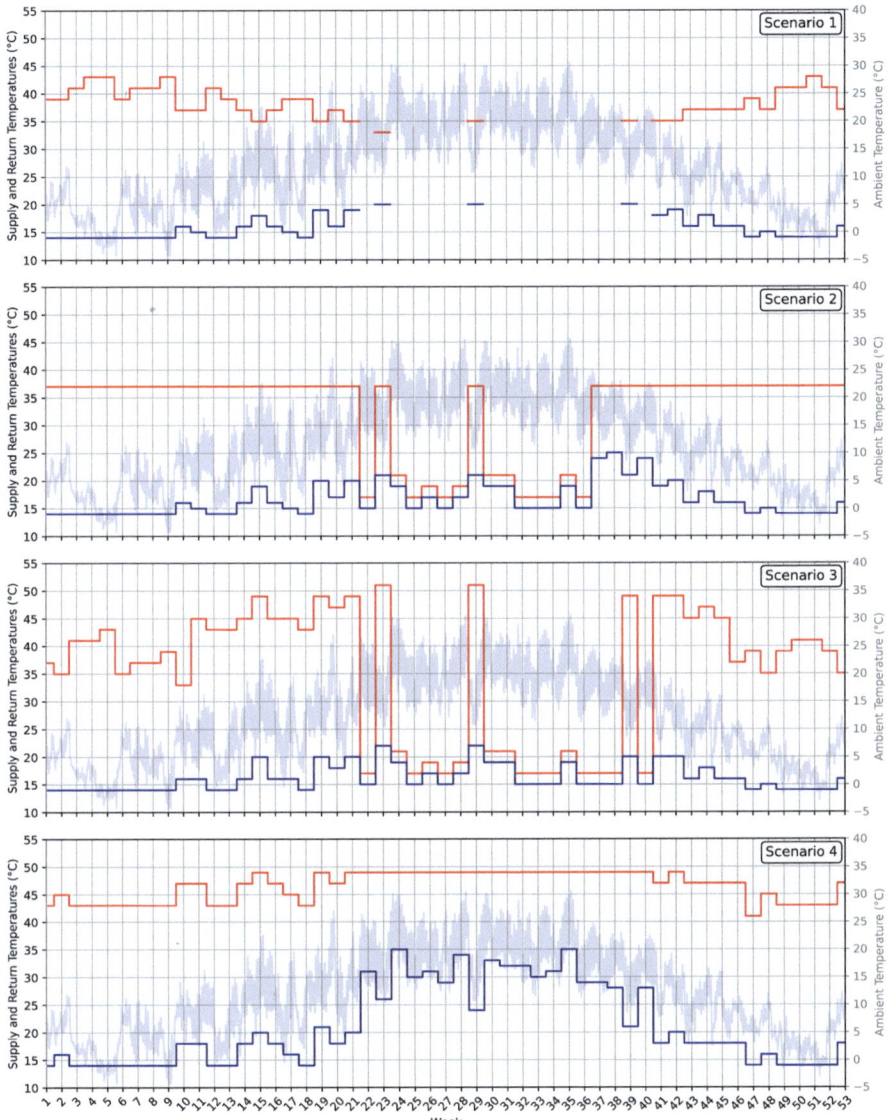

Fig. 3 District supply and return temperatures across weeks under different scenarios

Table 1 The total electricity consumption of the DHN across the different scenarios

Scenario	The total electricity consumption [MWh/year]
Scenario 1	910
Scenario 2	559
Scenario 3	520
Scenario 4	569

Figure 4 illustrates the duration load curves of the DHN's electricity load across all four scenarios. As expected, Scenario 1 shows the highest electricity load throughout the year, including the highest peak load. This is primarily due to the use of electric boilers for DHW production, which are less efficient than heat pumps. In Scenario 2, both the baseload and peak load are significantly reduced, thanks to decentralized heat pumps with high COPs. Scenario 3 demonstrates further slight improvement, with a consistently lower load curve than Scenario 2. This is attributed to the integration of booster heat pumps, which enable more efficient DHW production by offloading part of the demand from the main heat pump, thus further reducing the peak load. Finally, Scenario 4, despite incorporating thermal storage, shows slightly higher electricity consumption and peak load than Scenario 3. This is mainly because the storage system requires a higher temperature, which leads to increased thermal losses and higher energy input. It should also be noted that the storage parameters (e.g., charging/discharging schedules and tank sizes) are not optimized in this work, which may further influence the system's performance and efficiency.

It should be noted that the results and conclusions are valid within the context of low-temperature operation. In high-temperature systems, the choice of the most suitable scenario may differ, as the performance of each configuration can vary significantly due to changes in COP, thermal losses, and Legionella prevention requirements.

As a final remark, Scenarios 2 and 3 demonstrated the best performance in terms of both the electricity load profile and total energy consumption. While Scenario 2 performs slightly better energetically, it involves more equipment and decentralized heat pumps, which increases the overall system cost.

For future work, it would be valuable to explore the trade-offs between energy performance and economic aspects to identify the most suitable solution. Additionally, in the case of Scenario 4 with thermal storage, optimizing parameters such as charging/discharging schedules and storage tank sizes could provide further insights into their impact on the load curve and overall system efficiency.

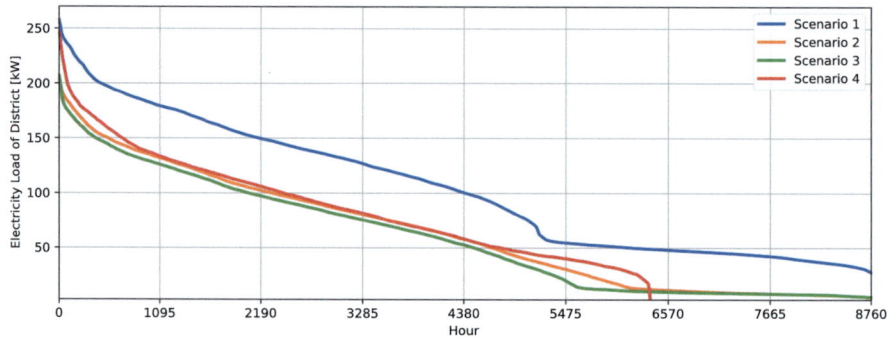

Fig. 4 Duration load curve of the district for all scenarios

4 Conclusion

This study evaluated four optimized scenarios for delivering space heating and domestic hot water in low-temperature district heating networks, aiming to minimize electricity consumption while ensuring comfort and hygiene. The analysis showed that using decentralized heat pumps for simultaneous SH and DHW production (Scenario 2) and integrating booster heat pumps for DHW (Scenario 3) significantly improve system efficiency compared to the other setups relying on electric boilers (Scenario 1) and thermal storage (Scenario 4). Scenario 3 achieved the lowest total electricity consumption and the most favorable load profile, highlighting the benefit of separating DHW generation via booster units. Overall, the findings underscore the importance of selecting DHW system configurations that effectively balance energy efficiency, operational flexibility, and compliance with health and safety standards. Future research should focus on techno-economic analysis, exploring trade-offs between energy savings and system investment costs.

Acknowledgements This published research was carried out with the support of the Swiss Federal Office of Energy (SFOE) as part of the SWEET DeCarbCH (Decarbonization of Cooling and Heating in Switzerland) under contract number Si/502260-01. The authors bear sole responsibility for the conclusions and the results presented in this publication.

References

1. IEA DHC, IEA DHC District heating network generation definitions. *iea-dhc.orgDH COOLINGiea-dhc.org*, Accessed Jul. 03, 2025. [Online]. Available: https://www.iea-dhc.org/fileadmin/public_documents/2402_IEA_DHC_DH_generations_definitions.pdf
2. Zhu T, Du Y, Liang J, Rohlfs W, Eric Thorsen J, Elmegaard B (2023) Integration of booster heat pumps in ultra-low temperature district heating network: prototype demonstration and refrigerant charge investigation. Energy Build 298:113516. https://doi.org/10.1016/J.ENBUILD.2023.113516
3. Bordignon S, Quaggiotto D, Vivian J, Emmi G, De Carli M, Zarrella A (2022) A solar-assisted low-temperature district heating and cooling network coupled with a ground-source heat pump. Energy Convers Manag 267:115838. https://doi.org/10.1016/J.ENCONMAN.2022.115838
4. Barbu M, Minciuc E, Frusescu DC, Tutica D (2021) Integration of hybrid photovoltaic thermal panels (PVT) in the district heating system of Bucharest, Romania. Proceedings of 2021 10th international conference on energy and environment, CIEM 2021. https://doi.org/10.1109/CIEM52821.2021.9614721
5. Legionella and heat pump water heaters. Accessed May 23, 2025. [Online]. Available: https://heatpumpingtechnologies.org
6. Chambers J, Cozza S, Patel M (2023) The GENEAP project: digitalising and automating planning of district heating and cooling. J Phys Conf Ser 2600(8):082003. https://doi.org/10.1088/1742-6596/2600/8/082003
7. Lidberg T, Olofsson T, Ödlund L (2019) Impact of domestic hot water systems on district heating temperatures. Energies 12(24):4694. https://doi.org/10.3390/EN12244694
8. Toffanin R, Curti V, Barbato MC (2021) Impact of Legionella regulation on a 4th generation district heating substation energy use and cost: the case of a Swiss single-family household. Energy 228:120473. https://doi.org/10.1016/J.ENERGY.2021.120473

Open Access This chapter is licensed under the terms of the Creative Commons Attribution 4.0 International License (http://creativecommons.org/licenses/by/4.0/), which permits use, sharing, adaptation, distribution and reproduction in any medium or format, as long as you give appropriate credit to the original author(s) and the source, provide a link to the Creative Commons license and indicate if changes were made.

The images or other third party material in this chapter are included in the chapter's Creative Commons license, unless indicated otherwise in a credit line to the material. If material is not included in the chapter's Creative Commons license and your intended use is not permitted by statutory regulation or exceeds the permitted use, you will need to obtain permission directly from the copyright holder.

Closing a Sim-to-Sim Gap for Automatic Fault Detection in DHC Systems Using Hybrid Modelling

Pieter Jan Houben, Stef Jacobs, Renzo Massobrio, Ivan Verhaert, and Peter Hellinckx

Abstract Accurate fault detection in district heating and cooling systems remains a challenge due to limitations in traditional simulation models. This paper explores a hybrid modelling approach that combines physics-based models with artificial neural networks to address this issue. Two methods of integrating these models are tested on simulated scenarios that replicate complex component behaviour. The first method applies corrections after full system simulation, while the second corrects outputs at the component level in real time. Both approaches reduce the discrepancy between simulated and reference data, but only the second method shows clear architectural changes based on component disturbances. This suggests a stronger potential for identifying faulty behaviour without the need for labelled datasets. The results demonstrate the value of integrating data-driven corrections at the component level to improve simulation accuracy and support fault detection in district heating and cooling systems.

Keywords HVAC simulation · Fault detection · Hybrid modelling · Neural networks

1 Introduction

Of global energy use, 40% is accounted for by buildings, of which 36% is used by their heating, ventilation and air-conditioning (HVAC) systems [6]. Moreover, due to undetected degradation and faults in components and improper controls, 15–30% of this energy use is wasted [1]. In district heating (DH) networks, inefficient building-level thermal management can lead to increased peak loads, higher return temperatures, and overall inefficiencies in heat distribution. These inefficiencies are

---

P. J. Houben (✉) · R. Massobrio · P. Hellinckx
M4S, Faculty of Applied Engineering - Electronics-ICT, University of Antwerp, Antwerp, Belgium
e-mail: pieterjan.houben@uantwerpen.be

S. Jacobs · I. Verhaert
EMIB, Faculty of Applied Engineering - Electronics-ICT, University of Antwerp, Antwerp, Belgium

caused by suboptimal control on the one hand and undetected faults or degradation of components on the other hand. Therefore, optimizing the connected building heating systems is crucial both for reducing energy use and for improving the performance and sustainability of DH networks. Optimizing control strategies and accurately simulating an HVAC plant are inevitably intertwined. That is, when components in a system show degradation and consequently behave differently, settings of the controller that were optimal before will not be optimal any more [2]. In the case of state-of-the-art methods like model predictive control (MPC) or deep reinforcement learning (DRL), this means that the underlying simulation model needs to accurately model the imperfections present in the designated HVAC system.

Current practice in HVAC simulation is the use of physics-based models (PBM), where thermodynamic formulas are simulated using timestep methods. Although these models simulate the system in a transparent way, they require simplifications to keep calculations feasible. This causes a discrepancy between the simulated behaviour and the actual behaviour, which is called the sim-to-real gap [8]. On the other hand, state-of-the-art research in HVAC simulation uses artificial neural networks (ANN) to simulate HVAC systems [5]. Although these models show great promise in terms of accuracy, they lack explainability, which makes it impossible to diagnose faults when the sim-to-real gap grows over time. That is why this paper aims to develop a methodology, combining PBMs and ANNs into a hybrid model that uses the ANNs to improve accuracy, while maintaining the explainability of the PBM, enabling automatic detection of faults and degradation. Two methods of combining PBMs and ANNs are compared in terms of accuracy and ability to detect deviant behaviour. To test this methodology, a sim-to-sim gap is closed instead of a sim-to-real gap, allowing reproduction of boundary conditions over different scenarios.

2 Methodology

2.1 Hybrid Modelling: Combining PBMs and ANNs

The goal of this paper is to test two different methods in which ANNs are used to correct the output of a PBM. In order to maintain the explainability of the PBM, the ANNs are added to the PBM at the level of the components.

2.1.1 Physics Based Model

The PBM that is used in this paper models the temperature T as a zero-order hold solution of Eq. 1 over a given timestep, where the values of a and b depend on the type of component that is being simulated. More information on the specifics for each component can be found in the work of Van Riet [7]. Furthermore, it uses occupant behaviour profiles generated by the Instal2020 project [9]. Weather profiles are based on data from Belgium between 2001 and 2020 [4]. This simulation model has already proven valuable in studies concerning efficiently controlling residential collective heating systems with decentralized storage tanks [3].

$$\frac{dT(t)}{dt} = -a(t)T(t) + b(t) \tag{1}$$

2.1.2 Artificial Neural Networks

An ANN consists of sub-models called neurons that are structured into layers that are connected to each other. Each neuron of the first layer takes a weighted sum of the input variables and then transforms this sum using a non-linear activation function. Next, each neuron of the second layer takes a weighted sum of the outputs of the neurons of the first layer, which is again transformed by a non-linear activation function. This process goes on until the neurons of the last layer produce the output variables. In this paper, each ANN has three layers and each layer has the same number of neurons as there are input variables. For each layer, a matrix can be constructed of which the columns are the weights of the neurons of that layer. Note that for a neuron in the first layer, the ith weight of that neuron represents the importance of the ith input variable for that neuron. Therefore, the ith row of the weight matrix of the first layer indicates the importance of the ith input variable in the model. That is why the weight matrix of the first layer will be used to represent the architecture of the ANNs. The following two paragraphs describe two methods to add ANNs to a PBM.

2.1.3 Method 1

The first method is shown in Fig. 1. It shows two components of the PBM, C_1 and C_2, as an example of how the calculations are performed within the model. More specifically, C_1 calculates its next state based only on its previous state, whereas C_2 uses output from C_1 as input. In this method, $S(t)$ is given to the PBM as input, which then calculates the next simulated state, denoted as $Sim(t+1)$. Next, the simulated state is corrected by an ANN, depicted as a black box, that outputs the next state, $S(t+1)$. This ANN takes as input $S(t)$ and $Sim(t)$. The white markings on the black box indicate that $Sim(t)$, as opposed to $S(t)$, is not passed in its entirety through the network, but that the network is divided per component. Therefore, the ANN is in fact a set of ANNs next to each other.

2.1.4 Method 2

The second method is shown in Fig. 2. The measured state is again given to the PBM. However, instead of calculating the entire simulated state $Sim(t)$, the output of each component is immediately corrected by a designated ANN. This means that

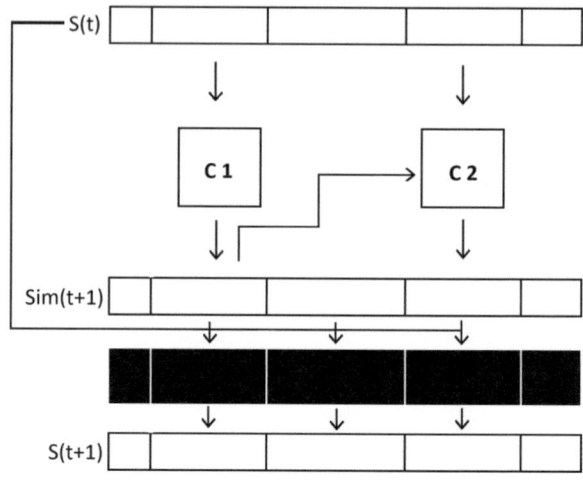

Fig. 1 First the PBM calculates the entire next simulated state, then one black box corrects this state

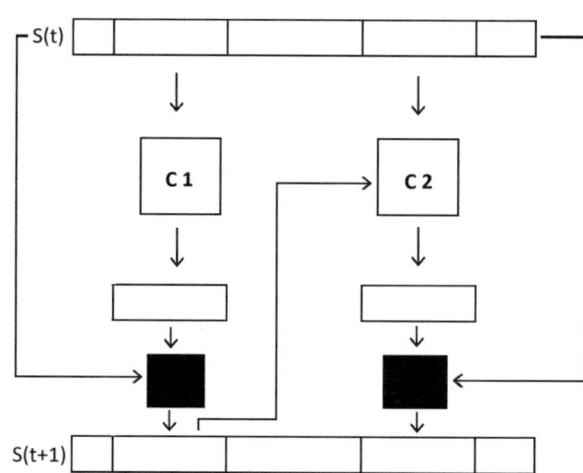

Fig. 2 The outputs of the components of the PBM are immediately corrected by the ANNs

the corrected output of C_1 is used as input for C_2 instead of the simulated output. As a consequence, the ANNs have to be trained simultaneously, since the output of one black box affects the input of the others. This makes the method less straightforward than the first.

In both methods, each ANN takes the entire state of the system together with the state of the component as predicted by the PBM as input to correct these predictions. This is done to allow the ANN to learn how the state of one component affects the state of another component.

2.2 Detecting Deviating Behaviour

When a component is behaving faulty, its behaviour will deviate more from the PBM than when it behaves perfectly. This paper aims to investigate how this is manifested in the weight matrix of the first layer of the ANN, allowing for a fault-detection mechanism that does not require data on faulty HVAC systems.

2.3 Sim-to-Sim Gap

The goal of this research is to compare the two methods presented in Sect. 2.1 on both accuracy and their ability to detect faults. Especially for this last part, it is important to perform experiments where different sets of components are behaving faulty. In order to be able to compare these experiments, all boundary conditions of the experiments should be the same. Therefore, instead of using actual sensor data, a dataset is simulated by adding a disturbance function to the output of the components in each timestep. This disturbance function mimics behaviour that is too complex to be simulated by a PBM and this dataset represents measured states of a system. Next, each state in this dataset is passed to the PBM, which then predicts the next state. Finally, the gap between the disturbed next state and the next state as predicted by the P is closed using both methods. This means that a sim-to-sim gap is closed.

The disturbance function $f(i)$ that is added to the output of the components in each timestep i is given by the following equation:

$$f(i) = A \cdot \left(2 + 100 \cdot \left(5\sin(\frac{i}{100}) + \zeta \cdot \cos\left(0.01\left(\frac{i}{100}\right)^2 + 0.1\frac{i}{100}\right)\right)\right),$$

where $\zeta = \frac{10}{1+\exp(-i/100+5)}$. This particular function is chosen for its complex behaviour on the one hand, and because its cumulative sum is bounded on the other. This last reason is important because an HVAC system is a slow-responding system, causing the disturbance function $f(i)$ to be added to the output many timesteps before the system can react to the disruption. If $\sum_i f(i)$ is not bounded, this can result in NaN values during the simulation, which is not desirable.

3 Results and Discussion

3.1 Case Study

To test the methods, a simple heating network is simulated in which 12 apartments are equipped with underfloor heating with design ingoing and outgoing temperature of 35 °C/30 °C. A geothermal heat pump provides heat with a heat power of 2.3 kW and is connected to a storage tank, designed to buffer 1 h of operation of the heat pump. The supply temperature towards the underfloor heating system is controlled by the central production valve, following a heating line. In the simulation, instead of modelling valves and pumps separately, desired flow rates are considered with a time delay. To create the training set resp. test set, this system was simulated over the course of a year for an elderly couple resp. couple of working-class people. Three scenarios were considered: one where only the heat pump was disturbed, one where only the storage tank was disturbed and one where both were disturbed. For each scenario, two hybrid models were created following each method.

3.2 Method 1

The first method succeeded in closing the sim-to-sim gap, measured using the mean squared error (MSE) on the output variables of the components, reducing it from values around 0.6 to the order of 0.01 on the test set. However, the weight matrices of the ANNs were very similar for each scenario, making it impossible to derive whether the components were being disturbed or not. This can be explained by the way the ANNs were added to the PBM. For example, in the scenario where the heat pump is disturbed, but the storage tank is not, the output of the storage tank still requires correction, as it depends on the uncorrected output of the heat pump to calculate its next simulated state.

3.3 Method 2

For the second method, the ANNs reduced the sim-to-sim gap from values around 0.6 to the order of 0.01, demonstrating their ability to correct the PBM. Moreover, their architectures differed significantly between the scenarios where their designated component was being disturbed or not. This is illustrated in Fig. 3. This figure shows the weight matrices of the model for the storage tank for each scenario. Figure 3a shows the matrix for the scenario where only the heat pump is disturbed, Fig. 3b and Fig. 3c show the matrices for the scenarios where the tank is disturbed. As explained in Sect. 2.1.2, all state variables as well as actions in the system are used as input for the ANNs and the rows of these matrices indicate the importance of their matching input variables, where values close to zero mean little importance. In these matrices darker colours indicate more importance. The dotted lines separate the rows that match output variables of different components. Knowing this, it can be observed that although three matrices look similar, the first matrix clearly emphasizes the variables having to do with the tank and its predictions over the other variables, whereas for the other two, the actions as well as portions of the other components are equally important. This indicates that method two is more appropriate to detect faults.

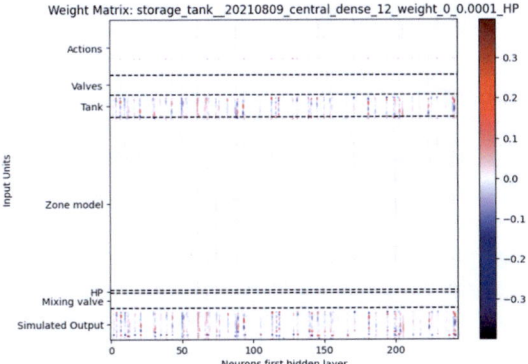

(a) Weight matrix of the first hidden layer for the model of the storage tank where only the heat pump was disturbed.

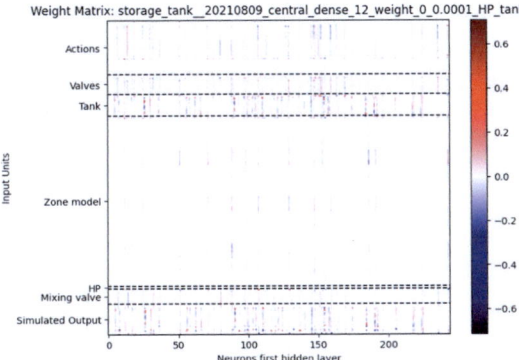

(b) Weight matrix where both the heat pump and the storage tank were disturbed.

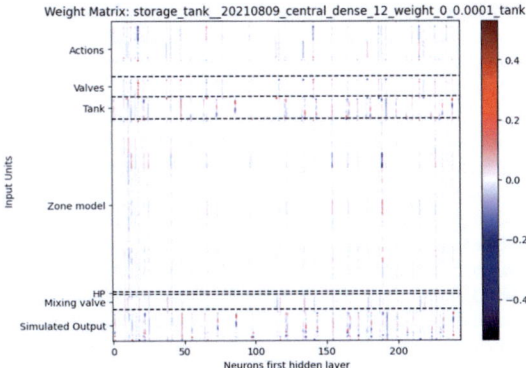

(c) Weight matrix where only the storage tank was disturbed.

Fig. 3 Weight matrices of the first hidden layer of the storage tank model across different disturbance scenarios

4 Conclusion and Future Work

This paper demonstrated the potential of hybrid modelling to improve the accuracy and fault detection capabilities of HVAC simulations within district heating and cooling systems. By integrating artificial neural networks with physics-based models, two hybrid approaches were evaluated. While both methods significantly reduced the sim-to-sim gap, the second method, where ANNs correct component outputs immediately, proved superior for identifying disturbances in components. These findings highlight the potential of hybrid models for improving simulation accuracy and enabling automatic fault detection without requiring labelled fault data, paving the way for more efficient and resilient thermal energy systems.

In future research, the second method will be further explored. For example, now it is assumed that all output variables of the components are measurable, allowing the ANNs to be trained using complete datasets. However, when working with a sim-to-real gap instead of a sim-to-sim gap, some output variables will not be measurable, meaning that more advanced techniques will have to be used to evaluate the ANNs both during and after training. Moreover, the findings of this research will have to be validated on different, possibly more complex systems. Finally, in order to perform actual automatic fault detection, specific criteria will have to be defined to automatically determine whether a component's behaviour is deviating or not.

Acknowledgements This research was supported by a Ph.D. fellowship of the Research Foundation Flanders (FWO) [1S54425N].

References

1. Chen Z, O'Neill Z, Wen J, Pradhan O, Yang T, Lu X, Lin G, Miyata S, Lee S, Shen C, Chiosa R, Piscitelli MS, Capozzoli A, Hengel F, Kührer A, Pritoni M, Liu W, Clauß J, Chen Y, Herr T (2023) A review of data-driven fault detection and diagnostics for building HVAC systems. Appl Energy 339:121030
2. Houben PJ, Jacobs S, Massobrio R, Tabari H, Verhaert I, Hellinckx P (2025) Evaluating the impact of suboptimal HVAC systems on control strategies. In: Barolli L (ed) Advances on P2P, Parallel, Grid, Cloud and Internet Computing. Springer Nature Switzerland, Cham, pp 347–355
3. Jacobs S, De Pauw M, Van Minnebruggen S, Ghane S, Huybrechts T, Hellinckx P, Verhaert I (2023) Grouped charging of decentralised storage to efficiently control collective heating systems: limitations and opportunities. Energies 16(8)
4. Klein S, Duffie J, Mithell J, Kummer J, Thornton J, Bradley D, Arias D, Beckman W, Duffieet N, Braun J (2009) Mathematical reference: type 54 (Hourly weather data generator)
5. Savadkoohi M, Macarulla M, Casals M (2023) Facilitating the implementation of neural network-based predictive control to optimize building heating operation. Energy 263:125703
6. UNEP (2021) Global status report for buildings and construction: towards a zero-emission, efficient and resilient buildings and construction sector. Technical report
7. Van Riet F (2019) Hydronic design of hybrid thermal production systems in buildings. University of Antwerp
8. Vera-Piazzini O, Scarpa M (2024) Building energy model calibration: a review of the state of the art in approaches, methods, and tools. J Build Eng 86:108287
9. Instal VLAIO (2020) Project: Integraal ontwerp van installaties voor sanitair en ver- warming (dutch). VIS 13589:2014–2018

Open Access This chapter is licensed under the terms of the Creative Commons Attribution 4.0 International License (http://creativecommons.org/licenses/by/4.0/), which permits use, sharing, adaptation, distribution and reproduction in any medium or format, as long as you give appropriate credit to the original author(s) and the source, provide a link to the Creative Commons license and indicate if changes were made.

The images or other third party material in this chapter are included in the chapter's Creative Commons license, unless indicated otherwise in a credit line to the material. If material is not included in the chapter's Creative Commons license and your intended use is not permitted by statutory regulation or exceeds the permitted use, you will need to obtain permission directly from the copyright holder.

MIX
Papier aus verantwortungsvollen Quellen
Paper from responsible sources
FSC® C105338

If you have any concerns about our products,
you can contact us on
ProductSafety@springernature.com

In case Publisher is established outside the EU,
the EU authorized representative is:
**Springer Nature Customer Service Center GmbH
Europaplatz 3, 69115 Heidelberg, Germany**

Printed by Libri Plureos GmbH
in Hamburg, Germany